Ben 710 (Ed.)

Perplicaria Boucheti

Ben 710 (Ed.)

Perplicaria Boucheti

Sea Snail, Mollusca, Cancellariidae

Part Press

Publisher:
Part Press is a trademark of
International Book Market Service Ltd., 17 Rue Meldrum, Beau Bassin, 1713-01 Mauritius
Email: info@bookmarketservice.com
Website: www.bookmarketservice.com

Published in 2011

Printed in: U.S.A., U.K., Germany. This book was not produced in Mauritius.

ISBN: 978-613-8-91731-1

Contents

Perplicaria boucheti

Perplicaria boucheti	
Scientific classification	
Kingdom:	Animalia
Phylum:	Mollusca
Class:	Gastropoda
Unranked:	clade Caenogastropoda clade Hypsogastropoda clade Neogastropoda
Superfamily:	Cancellarioidea
Family:	Cancellariidae
Genus:	*Perplicaria*
Species:	*P. boucheti*
Binomial name	
Perplicaria boucheti Verhecken, 1997	

Perplicaria boucheti is a species of sea snail, a marine gastropod mollusk in the family Cancellariidae, the nutmeg snails.[1]

References

[1] *Perplicaria boucheti* Verhecken, 1997 (http://www.marinespecies.org/aphia.php?p=taxdetails&id=464801). Accessed through: World Register of Marine Species at http://www.marinespecies.org/aphia.php?p=taxdetails&id=464801 on 6 April 2010.

Perplicaria

Perplicaria	
Scientific classification	
Kingdom:	Animalia
Phylum:	Mollusca
Class:	Gastropoda
Unranked:	clade Caenogastropoda clade Hypsogastropoda clade Neogastropoda
Superfamily:	Cancellarioidea
Family:	Cancellariidae
Genus:	*Perplicaria* Dall, 1890

Perplicaria is a genus of sea snails, marine gastropod mollusks in the family Cancellariidae, the nutmeg snails.[1]

Species

Species within the genus *Perplicaria* include:

- *Perplicaria boucheti* Verhecken, 1997[2]
- *Perplicaria clarki* M. Smith, 1947[3]

References

[1] *Perplicaria* Dall, 1890 (http://www.marinespecies.org/aphia.php?p=taxdetails&id=464485). Accessed through: World Register of Marine Species at http://www.marinespecies.org/aphia.php?p=taxdetails&id=464485 on 13 April 2010.

[2] *Perplicaria boucheti* Verhecken, 1997 (http://www.marinespecies.org/aphia.php?p=taxdetails&id=464801). Accessed through: World Register of Marine Species at http://www.marinespecies.org/aphia.php?p=taxdetails&id=464801 on 13 April 2010.

[3] *Perplicaria clarki* M. Smith, 1947 (http://www.marinespecies.org/aphia.php?p=taxdetails&id=464802). Accessed through: World Register of Marine Species at http://www.marinespecies.org/aphia.php?p=taxdetails&id=464802 on 13 April 2010.

Species

In biology, a **species** is one of the basic units of biological classification and a taxonomic rank. A species is often defined as a group of organisms capable of interbreeding and producing fertile offspring. While in many cases this definition is adequate, more precise or differing measures are often used, such as similarity of DNA, morphology or ecological niche. Presence of specific locally adapted traits may further subdivide species into subspecies.

Species that are believed to have the same ancestors are grouped together, and this group is called a genus. A species can only belong to one genus that it was grouped into. The belief is best checked by a similarity of their DNA, but for practical reasons, other similar properties are used. For plants similarities of flowers are used. All species are given a two part name (a "binomial name"). The first part of a binomial name is the generic name, the genus of the species. The second part is either the specific name (a term used only in zoology, never in botany, for the second part of a binomial) or the specific epithet (the term always used in botany, which can also be used in zoology). For example, *Boa constrictor*, which is commonly called by its binomial name, and is one of four species of the *Boa* genus. The first part of the name is capitalized, and the second part has a lower case. The two part name is written in italics.

A usable definition of the word "species" and reliable methods of identifying particular species are essential for stating and testing biological theories and for measuring biodiversity, though other taxonomic levels such as families may be considered in broad scale studies.[1] Extinct species known only from fossils are generally difficult to assign precise taxonomic rankings, which is why higher taxonomic levels such as families are often used for fossil based studies.[1] [2]

The total number of non-bacterial species in the world has been estimated at 8.7 million,[3] with previous estimates ranging from two million to 100 million.[4]

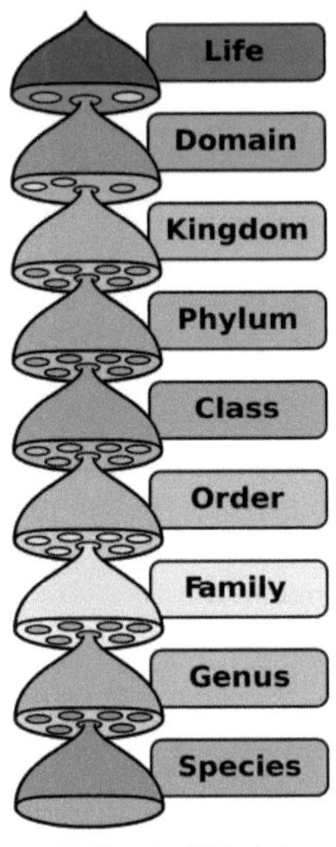

The hierarchy of biological classification's eight major taxonomic ranks, which is an example of definition by genus and differentia. A genus contains one or more species. Intermediate minor rankings are not shown.

Biologists' working definition

A usable definition of the word "species" and reliable methods of identifying particular species is essential for stating and testing biological theories and for measuring biodiversity. Traditionally, multiple examples of a proposed species must be studied for unifying characters before it can be regarded as a species. It is generally difficult to give precise taxonomic rankings to extinct species known only from fossils.

Some biologists may view species as statistical phenomena, as opposed to the traditional idea, with a species seen as a class of organisms. In that case, a species is defined as a separately evolving lineage that forms a single gene pool. Although properties such as DNA-sequences and morphology are used to help separate closely related lineages,[5] this definition has fuzzy boundaries.[6] However, the exact definition of the term "species" is still controversial, particularly in prokaryotes,[7] and this is called the species problem.[8] Biologists have proposed a range of more precise definitions, but the definition used is a pragmatic choice that depends on the particularities of the species of concern.[8]

Common names and species

The commonly used names for plant and animal taxa sometimes correspond to species: for example, "lion", "walrus", and "Camphor tree" – each refers to a species. In other cases common names do not: for example, "deer" refers to a family of 34 species, including Eld's Deer, Red Deer and Elk (Wapiti). The last two species were once considered a single species, illustrating how species boundaries may change with increased scientific knowledge.

Placement within genera

Ideally, a species is given a formal, scientific name, although in practice there are very many unnamed species (which have only been described, not named). When a species is named, it is placed within a genus. From a scientific point of view this can be regarded as a hypothesis that the species is more closely related to other species within its genus (if any) than to species of other genera. Species and genus are usually defined as part of a larger taxonomic hierarchy. The best-known taxonomic ranks are, in order: life, domain, kingdom, phylum, class, order, family, genus, and species. This assignment to a genus is not immutable; later a different (or the same) taxonomist may assign it to a different genus, in which case the name will also change.

In biological nomenclature, the name for a species is a two-part name (a binomial name), treated as Latin, although roots from any language can be used as well as names of locales or individuals. The generic name is listed first (with its leading letter capitalized), followed by a second term. The terminology used for the second term differs between zoological and botanical nomenclature.

- In zoological nomenclature, the second part of the name can be called the specific name or the specific epithet. For example, gray wolves belong to the species *Canis lupus*, coyotes to *Canis latrans*, golden jackals to *Canis aureus*, etc., and all of those belong to the genus *Canis* (which also contains many other species). For the gray wolf, the genus name is *Canis*, the specific name or specific epithet is *lupus*, and the binomen, the name of the species, is *Canis lupus*.
- In botanical nomenclature, the second part of the name can only be called the specific epithet. The 'specific name' in botany is always the combination of genus name and specific epithet. For example, the species commonly known as the longleaf pine is *Pinus palustris*; the genus name is *Pinus*, the specific epithet is *palustris*, the specific name is *Pinus palustris*.

This binomial naming convention, later formalized in the biological codes of nomenclature, was first used by Leonhart Fuchs and introduced as the standard by Carolus Linnaeus in his 1753, *Species Plantarum* (followed by his, 1758 *Systema Naturae*, 10th edition). At that time, the chief biological theory was that species represented independent acts of creation by God and were therefore considered objectively real and immutable, so the hypothesis of common descent did not apply.

Abbreviated names

Books and articles sometimes intentionally do not identify species fully and use the abbreviation "**sp.**" in the singular or "**spp.**" in the plural in place of the specific epithet: for example, ***Canis* sp.** This commonly occurs in the following types of situations:

- The authors are confident that some individuals belong to a particular genus but are not sure to which exact species they belong. This is particularly common in paleontology.
- The authors use "spp." as a short way of saying that something applies to many species within a genus, but do not wish to say that it applies to all species within that genus. If scientists mean that something applies to all species within a genus, they use the genus name without the specific epithet.

In books and articles, genus and species names are usually printed in italics. If using "sp." and "spp.", these should not be italicized.

Identification codes

Various codes have been devised for identifying particular species. For example:

- NCBI employs a numeric 'taxid' or *Taxonomy identifier*, a "stable unique identifier", e.g. the taxid of *H. sapiens* is 9606 [9]
- KEGG employs a three-letter code for a limited number of organisms; in this code, for example, *H. sapiens* is simply *hsa* [10];
- UniProt employs an "organism mnemonic" of not more than five alphanumeric characters, e.g. *HUMAN* for *H. sapiens* [11]

Difficulty of defining "species" and identifying particular species

It is surprisingly difficult to define the word "species" in a way that applies to all naturally occurring organisms, and the debate among biologists about how to define "species" and how to identify actual species is called the species problem. Over two dozen distinct definitions of "species" are in use amongst biologists.[12]

Most textbooks follow Ernst Mayr's definition of a species as "groups of actually or potentially interbreeding natural populations, which are reproductively isolated from other such groups".[8]

The Greenish Warbler demonstrates the concept of a ring species.

Various parts of this definition serve to exclude some unusual or artificial matings:

- Those that occur only in captivity (when the animal's normal mating partners may not be available) or as a result of deliberate human action
- Animals that may be physically and physiologically capable of mating but, for various reasons, do not normally do so in the wild

The typical textbook definition above works well for most multi-celled organisms, but there are several types of situations in which it breaks down:

- By definition it applies only to organisms that reproduce sexually. So it does not work for asexually reproducing single-celled organisms and for the relatively few parthenogenetic multi-celled organisms. The term "phylotype" is often applied to such organisms.
- Biologists frequently do not know whether two morphologically similar groups of organisms are "potentially" capable of interbreeding.
- There is considerable variation in the degree to which hybridization may succeed under natural conditions, or even in the degree to which some organisms use sexual reproduction between individuals to breed.
- In ring species, members of adjacent populations interbreed successfully but members of some non-adjacent populations do not.
- In a few cases it may be physically impossible for animals that are members of the same species to mate. However, these are cases in which human intervention has caused gross morphological changes, and are therefore excluded by the biological species concept.

Horizontal gene transfer makes it even more difficult to define the word "species". There is strong evidence of horizontal gene transfer between very dissimilar groups of prokaryotes, and at least occasionally between dissimilar groups of eukaryotes; and Williamson[13] argues that there is evidence for it in some crustaceans and echinoderms. All definitions of the word "species" assume that an organism gets all its genes from one or two parents that are very like that organism, but horizontal gene transfer makes that assumption false.

Definitions of species

The question of how best to define "species" is one that has occupied biologists for centuries, and the debate itself has become known as the species problem. Darwin wrote in chapter II of *On the Origin of Species*:

> No one definition has satisfied all naturalists; yet every naturalist knows vaguely what he means when he speaks of a species. Generally the term includes the unknown element of a distinct act of creation.[14]

But later, in *The Descent of Man*, when addressing "The question whether mankind consists of one or several species", Darwin revised his opinion to say:

> it is a hopeless endeavour to decide this point on sound grounds, until some definition of the term "species" is generally accepted; and the definition must not include an element that cannot possibly be ascertained, such as an act of creation.[15]

The modern theory of evolution depends on a fundamental redefinition of "species". Prior to Darwin, naturalists viewed species as ideal or general types, which could be exemplified by an ideal specimen bearing all the traits general to the species. Darwin's theories shifted attention from uniformity to variation and from the general to the particular. According to intellectual historian Louis Menand,

> Once our attention is redirected to the individual, we need another way of making generalizations. We are no longer interested in the conformity of an individual to an ideal type; we are now interested in the relation of an individual to the other individuals with which it interacts. To generalize about groups of interacting individuals, we need to drop the language of types and essences, which is prescriptive (telling us what finches should be), and adopt the language of statistics and probability, which is predictive (telling us what the average finch, under specified conditions, is likely to do). Relations will be more important than categories; functions, which are variable, will be more important than purposes; transitions will be more important than boundaries; sequences will be more important than hierarchies.

This shift results in a new approach to "species"; Darwin

> concluded that species are what they appear to be: ideas, which are provisionally useful for naming groups of interacting individuals. "I look at the term species", he wrote, "as one arbitrarily given for the sake of convenience to a set of individuals closely resembling each other ... It does not essentially differ from the word variety, which is given to less distinct and more fluctuating forms. The term variety, again, in comparison with mere individual differences, is also applied arbitrarily, and for convenience sake." [16]

Practically, biologists define species as *populations of organisms that have a high level of genetic similarity*. This may reflect an adaptation to the same niche, and the transfer of genetic material from one individual to others, through a variety of possible means. The exact level of similarity used in such a definition is arbitrary, but this is the most common definition used for organisms that reproduce asexually (asexual reproduction), such as some plants and microorganisms.

This lack of any clear species concept in microbiology has led to some authors arguing that the term "species" is not useful when studying bacterial evolution. Instead they see genes as moving freely between even distantly related bacteria, with the entire bacterial domain being a single gene pool. Nevertheless, a kind of rule of thumb has been established, saying that species of *Bacteria* or *Archaea* with 16S rRNA gene sequences more similar than 97% to each other need to be checked by DNA-DNA Hybridization if they belong to the same species or not.[17] This concept has been updated recently, saying that the border of 97% was too low and can be raised to 98.7%.[18]

In the study of sexually reproducing organisms, where genetic material is shared through the process of reproduction, the ability of two organisms to interbreed and produce fertile offspring of both sexes is generally accepted as a simple indicator that the organisms share enough genes to be considered members of the same species. Thus a "species" is a group of interbreeding organisms.

This definition can be extended to say that a species is a group of organisms that could potentially interbreed – fish could still be classed as the same species even if they live in different lakes, as long as they could still interbreed were they ever to come into contact with each other. On the other hand, there are many examples of series of three or more distinct populations, where individuals of the population in the middle can interbreed with the populations to either side, but individuals of the populations on either side cannot interbreed. Thus, one could argue that these populations constitute a single species, or two distinct species. This is not a paradox; it is evidence that species are defined by gene frequencies, and thus have fuzzy boundaries.

Consequently, any single, universal definition of "species" is necessarily arbitrary. Instead, biologists have proposed a range of definitions; which definition a biologists uses is a pragmatic choice, depending on the particularities of that biologist's research.

In practice, these definitions often coincide, and the differences between them are more a matter of emphasis than of outright contradiction. Nevertheless, no species concept yet proposed is entirely objective, or can be applied in all cases without resorting to judgment. Given the complexity of life, some have argued that such an objective definition is in all likelihood impossible, and biologists should settle for the most practical definition.

For most vertebrates, this is the biological species concept (BSC), and to a lesser extent (or for different purposes) the phylogenetic species concept (PSC). Many BSC subspecies are considered species under the PSC; the difference between the BSC and the PSC can be summed up insofar as that the BSC defines a species as a consequence of manifest evolutionary *history*, while the PSC defines a species as a consequence of manifest evolutionary *potential*. Thus, a PSC species is "made" as soon as an evolutionary lineage has started to separate, while a BSC species starts to exist only when the lineage separation is complete. Accordingly, there can be considerable conflict between alternative classifications based upon the PSC versus BSC, as they differ completely in their treatment of taxa that would be considered subspecies under the latter model (e.g., the numerous subspecies of honey bees).

Typological species

A group of organisms in which individuals are members of the species if they sufficiently conform to certain fixed properties or "rights of passage". The clusters of variations or phenotypes within specimens (i.e. longer or shorter tails) would differentiate the species. This method was used as a "classical" method of determining species, such as with Linnaeus early in evolutionary theory. However, we now know that different phenotypes do not always constitute different species (e.g.: a 4-winged Drosophila born to a 2-winged mother is not a different species). Species named in this manner are called *morphospecies*.[19]

Evolutionary species

A single evolutionary lineage of organisms within which genes can be shared, and that maintains its integrity with respect to other lineages through both time and space. At some point in the evolution of such a group, some members may diverge from the main population and evolve into a subspecies, a process that may eventually lead to the formation of a new species if isolation (geographical or ecological) is maintained. A species that gives rise to another species is a paraphyletic species, or paraspecies. [20] [21]

Phylogenetic (Cladistic)

A group of organisms that shares an ancestor; a lineage that maintains its integrity with respect to other lineages through both time and space. At some point in the progress of such a group, members may diverge from one another: when such a divergence becomes sufficiently clear, the two populations are regarded as separate species. This differs from evolutionary species in that the parent species goes extinct taxonomically when a new species evolve, the mother and daughter populations now forming two new species.[22] Subspecies as such are not recognized under this approach; either a population is a phylogenetic species or it is not taxonomically distinguishable.

Other

Ecological species

> A set of organisms adapted to a particular set of resources, called a niche, in the environment. According to this concept, populations form the discrete phenetic clusters that we recognize as species because the ecological and evolutionary processes controlling how resources are divided up tend to produce those clusters.

Biological / reproductive species

> Two organisms that are able to reproduce naturally to produce fertile offspring of both sexes. Organisms that can reproduce but almost always make infertile hybrids of at least one sex, such as a mule, hinny or F1 male cattalo are not considered to be the same species.

Biological / Isolation species

> A set of actually or potentially interbreeding populations. This is generally a useful formulation for scientists working with living examples of the higher taxa like mammals, fish, and birds, but more problematic for organisms that do not reproduce sexually. The results of breeding experiments done in artificial conditions may or may not reflect what would happen if the same organisms encountered each other in the wild, making it difficult to gauge whether or not the results of such experiments are meaningful in reference to natural populations.

Genetic species

> Based on similarity of DNA of individuals or populations. Techniques to compare similarity of DNA include DNA-DNA hybridization, and genetic fingerprinting (or DNA barcoding).

Cohesion species

> Most inclusive population of individuals having the potential for phenotypic cohesion through intrinsic cohesion mechanisms. This is an expansion of the mate-recognition species concept to allow for post-mating isolation mechanisms; no matter whether populations can hybridize successfully, they are still distinct cohesion species if the amount of hybridization is insufficient to completely mix their respective gene pools.

Evolutionarily Significant Unit (ESU)

> An evolutionarily significant unit is a population of organisms that is considered distinct for purposes of conservation. Often referred to as a species or a *wildlife species*, an ESU also has several possible definitions, which coincide with definitions of species.

Morphological species

> A population or group of populations that differs morphologically from other populations. For example, we can distinguish between a chicken and a duck because they have different shaped bills and the duck has webbed feet. Species have been defined in this way since well before the beginning of recorded history. This species concept is highly criticized because more recent genetic data reveal that genetically distinct populations may look very similar and, contrarily, large morphological differences sometimes exist between very closely related populations. Nonetheless, most species known have been described solely from morphology.

Phenetic species

> Based on phenotypes.

Microspecies

> Species that reproduce without meiosis or fertilization so that each generation is genetically identical to the previous generation. See also apomixis.

Recognition species

Based on shared reproductive systems, including mating behavior. The Recognition concept of species has been introduced by Hugh E. H. Paterson, after earlier work by Wilhelm Petersen.

Mate-recognition species

A group of organisms that are known to recognize one another as potential mates. Like the isolation species concept above, it applies only to organisms that reproduce sexually. Unlike the isolation species concept, it focuses specifically on pre-mating reproductive isolation.

Numbers of species

Bearing in mind the aforementioned problems with categorising species, the following numbers are only a soft guide. In 2010, they broke down as follows:[23]

Total number of species (estimated): 7–100 million (identified and unidentified), including:

- 5–10 million bacteria;[24]
- 1.5 million fungi;[25]
- ~1 million mites[26]
- 10–30 million insects;[27]

Number of *identified* eukaryote species: 1.6 million, including:[28]

- 3,067 brown algae
- 321,212 plants, including:
  - 10,134 red & green algae
  - 16,236 mosses,
  - 12,000 ferns & horsetails,
  - 1,021 gymnosperms,
  - 281,821 angiosperms;
- 74,000-120,000 fungi;[25]
- 17,000 lichens;
- 1,367,555 animals, including:
  - 1,305,250 invertebrates
    - 2,175 corals
    - 85,000 mollusks
    - 102,248 arachnids
    - 47,000 crustaceans
    - 1,000,000 insects
    - 68,827 other invertebrates;
  - 62,305 vertebrates
    - 31,300 fish,
    - 6,433 amphibians,
    - 9,084 reptiles,
    - 9,998 birds,

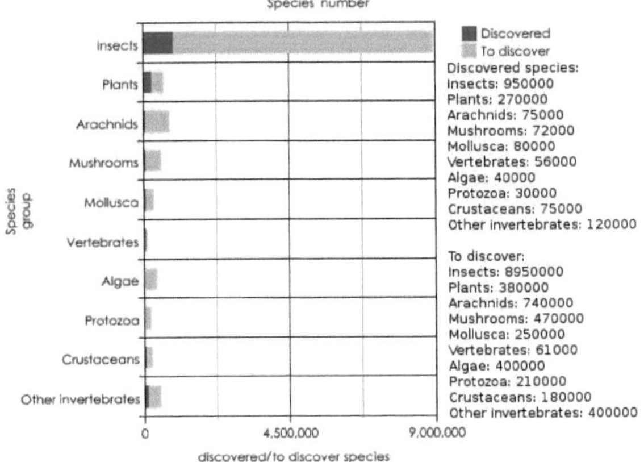

Undiscovered and discovered species

- 5,490 mammals;

At present, organisations such as the Global Taxonomy Initiative, the European Distributed Institute of Taxonomy and the Census of Marine Life[29] (the latter only for marine organisms) are trying to improve taxonomy and implement previously undiscovered species to the taxonomy system. Because we know but a portion of the organisms in the biosphere, we do not have a complete understanding of the workings of our environment. To make matters worse, despite the discovery of new species, according to professor James Mallet, we are wiping out these species at an unprecedented rate.[30] This means that even before a new species has had the chance of being studied and classified, it may already be extinct.

Importance in biological classification

The idea of *species* has a long history. It is one of the most important levels of classification, for several reasons:

- It often corresponds to what lay people treat as the different basic kinds of organism – dogs are one species, cats another.
- It is the standard binomial nomenclature (or trinomial nomenclature) by which scientists typically refer to organisms.
- It is the highest taxonomic level that cannot be made more or less inclusionary.

After years of use, the concept remains central to biology and a host of related fields, and yet also remains at times ill-defined.

Implications of assignment of species status

The naming of a particular species may be regarded as a *hypothesis* about the evolutionary relationships and distinguishability of that group of organisms. As further information comes to hand, the hypothesis may be confirmed or refuted. Sometimes, especially in the past when communication was more difficult, taxonomists working in isolation have given two distinct names to individual organisms later identified as the same species. When two named species are discovered to be of the same species, the older species name is usually retained, and the newer species name dropped, a process called *synonymization*, or colloquially, as **lumping**. Dividing a taxon into multiple, often new, taxons is called **splitting**. Taxonomists are often referred to as "lumpers" or "splitters" by their colleagues, depending on their personal approach to recognizing differences or commonalities between organisms (see lumpers and splitters).

Traditionally, researchers relied on observations of anatomical differences, and on observations of whether different populations were able to interbreed successfully, to distinguish species; both anatomy and breeding behavior are still important to assigning species status. As a result of the revolutionary (and still ongoing) advance in microbiological research techniques, including DNA analysis, in the last few decades, a great deal of additional knowledge about the differences and similarities between species has become available. Many populations formerly regarded as separate species are now considered a single taxon, and many formerly grouped populations have been split. Any taxonomic level (species, genus, family, etc.) can be synonymized or split, and at higher taxonomic levels, these revisions have been still more profound.

From a taxonomical point of view, groups within a species can be defined as being of a taxon hierarchically lower than a species. In zoology only the subspecies is used, while in botany the variety, subvariety, and form are used as well. In conservation biology, the concept of evolutionary significant units (ESU) is used, which may define either species or smaller distinct population segments. Identifying and naming species is the providence of alpha taxonomy.

Historical development of the species concept

In the earliest works of science, a species was simply an individual organism that represented a group of similar or nearly identical organisms. No other relationships beyond that group were implied. Aristotle used the words *genus* and *species* to mean generic and specific categories. Aristotle and other pre-Darwinian scientists took the species to be distinct and unchanging, with an "essence", like the chemical elements. When early observers began to develop systems of organization for living things, they began to place formerly isolated species into a context. Many of these early delineation schemes would now be considered whimsical and these included consanguinity based on color (all plants with yellow flowers) or behavior (snakes, scorpions and certain biting ants).

In the 18th century Swedish scientist Carolus Linnaeus classified organisms according to differences in the form of reproductive apparatus. Although his system of classification sorts organisms

Linnaeus believed in the fixity of species.

according to degrees of similarity, it made no claims about the relationship between similar species. At that time, it was still widely believed that there was no organic connection between species, no matter how similar they appeared. This approach also suggested a type of idealism: the notion that each species existed as an "ideal form". Although there are always differences (although sometimes minute) between individual organisms, Linnaeus considered such variation problematic. He strove to identify individual organisms that were exemplary of the species, and considered other non-exemplary organisms to be deviant and imperfect.

By the 19th century most naturalists understood that species could change form over time, and that the history of the planet provided enough time for major changes. Jean-Baptiste Lamarck, in his 1809 *Zoological Philosophy*, offered one of the first logical arguments against creationism. The new emphasis was on determining *how* a species could change over time. Lamarck suggested that an organism could pass on an acquired trait to its offspring, i.e., the giraffe's long neck was attributed to generations of giraffes stretching to reach the leaves of higher treetops (this well-known and simplistic example, however, does not do justice to the breadth and subtlety of Lamarck's ideas). With the acceptance of the natural selection idea of Charles Darwin in the 1860s, however, Lamarck's view of goal-oriented evolution, also known as a teleological process, was eclipsed. Recent interest in inheritance of acquired characteristics centers around epigenetic processes, e.g. methylation, that do not affect DNA sequences, but instead alter expression in an inheritable manner. Thus, neo-lamarckism, as it is sometimes termed, is not a challenge to the theory of evolution by natural selection.

Charles Darwin and Alfred Wallace provided what scientists now consider as the most powerful and compelling theory of evolution. Darwin argued that it was populations that evolved, not individuals. His argument relied on a radical shift in perspective from that of Linnaeus: rather than defining species in ideal terms (and searching for an ideal representative and rejecting deviations), Darwin considered variation among individuals to be natural. He further argued that variation, far from being problematic, actually provides the *explanation* for the existence of distinct species.

Darwin's work drew on Thomas Malthus' insight that the rate of growth of a biological population will always outpace the rate of growth of the resources in the environment, such as the food supply. As a result, Darwin argued, not all the members of a population will be able to survive and reproduce. Those that did will, on average, be the ones possessing variations—however slight—that make them slightly better adapted to the environment. If these variable traits are heritable, then the offspring of the survivors will also possess them. Thus, over many generations, adaptive variations will accumulate in the population, while counter-adaptive traits will tend to be eliminated.

Whether a variation is adaptive or non-adaptive depends on the environment: different environments favor different traits. Since the environment effectively selects which organisms live to reproduce, it is the environment (the "fight for existence") that selects the traits to be passed on. This is the theory of evolution by natural selection. In this model, the length of a giraffe's neck would be explained by positing that proto-giraffes with longer necks would have had a significant reproductive advantage to those with shorter necks. Over many generations, the entire population would be a species of long-necked animals.

In 1859, when Darwin published his theory of natural selection, the mechanism behind the inheritance of individual traits was unknown. Although Darwin made some speculations on how traits are inherited (pangenesis), his theory relies only on the fact that inheritable traits *exist*, and are variable (which makes his accomplishment even more remarkable.) Although Gregor Mendel's paper on genetics was published in 1866, its significance was not recognized. It was not until 1900 that his work was rediscovered by Hugo de Vries, Carl Correns and Erich von Tschermak, who realised that the "inheritable traits" in Darwin's theory are genes.

The theory of the evolution of species through natural selection has two important implications for discussions of species—consequences that fundamentally challenge the assumptions behind Linnaeus' taxonomy. First, it suggests that species are not just similar, they may actually be related. Some students of Darwin argue that *all* species are descended from a common ancestor. Second, it supposes that "species" are not homogeneous, fixed, permanent things; members of a species are all different, and over time species change. This suggests that species do not have any clear boundaries but are rather momentary statistical effects of constantly changing gene-frequencies. One may still use Linnaeus' taxonomy to identify individual plants and animals, but one can no longer think of species as independent and immutable.

The rise of a new species from a parental line is called speciation. There is no clear line demarcating the ancestral species from the descendant species.

Although the current scientific understanding of species suggests that there is no rigorous and comprehensive way to distinguish between different species in *all* cases, biologists continue to seek concrete ways to operationalize the idea. One of the most popular biological definitions of species is in terms of reproductive isolation; if two creatures cannot reproduce to produce fertile offspring of both sexes, then they are in different species. This definition captures a number of intuitive species boundaries, but it remains imperfect. It has nothing to say about species that reproduce asexually, for example, and it is very difficult to apply to extinct species. Moreover, boundaries between species are often fuzzy: there are examples where members of one population can produce fertile offspring of both sexes with a second population, and members of the second population can produce fertile offspring of both sexes with members of a third population, but members of the first and third population cannot produce fertile offspring, or can only produce fertile offspring of the homozygous sex. Consequently, some people reject this definition of a species.

Richard Dawkins defines two organisms as conspecific if and only if they have the same number of chromosomes and, for each chromosome, both organisms have the same number of nucleotides (*The Blind Watchmaker*, p. 118). However, most if not all taxonomists would strongly disagree. For example, in many amphibians, most notably in New Zealand's *Leiopelma* frogs, the genome consists of "core" chromosomes that are mostly invariable and accessory chromosomes, of which exist a number of possible combinations. Even though the chromosome numbers are highly variable between populations, these can interbreed successfully and form a single evolutionary unit. In plants, polyploidy is extremely commonplace with few restrictions on interbreeding; as individuals with an odd number of chromosome sets are usually sterile, depending on the actual number of chromosome sets present, this results in the odd situation where some individuals of the same evolutionary unit can interbreed with certain others and some cannot, with all populations being eventually linked as to form a common gene pool.

The classification of species has been profoundly affected by technological advances that have allowed researchers to determine relatedness based on molecular markers, starting with the comparatively crude blood plasma precipitation assays in the mid-20th century to Charles Sibley's ground-breaking DNA-DNA hybridization studies in the 1970s leading to DNA sequencing techniques. The results of these techniques caused revolutionary changes in the higher taxonomic categories (such as phyla and classes), resulting in the reordering of many branches of the phylogenetic tree (*see also:* molecular phylogeny). For taxonomic categories below genera, the results have been mixed so far; the pace of evolutionary change on the molecular level is rather slow, yielding clear differences only after considerable periods of reproductive separation. DNA-DNA hybridization results have led to misleading conclusions, the Pomarine Skua – Great Skua phenomenon being a famous example. Turtles have been determined to evolve with just one-eighth of the speed of other reptiles on the molecular level, and the rate of molecular evolution in albatrosses is half of what is found in the rather closely related storm-petrels. The hybridization technique is now obsolete and is replaced by more reliable computational approaches for sequence comparison. Molecular taxonomy is not directly based on the evolutionary processes, but rather on the overall change brought upon by these processes. The processes that lead to the generation and maintenance of variation such as mutation, crossover and selection are not uniform (see also molecular clock). DNA is only extremely rarely a direct target of natural selection rather than changes in the DNA sequence enduring over generations being a result of the latter; for example, silent transition-transversion combinations would alter the melting point of the DNA sequence, but not the sequence of the encoded proteins and thus are a possible example where, for example in microorganisms, a mutation confers a change in fitness all by itself.

See also

- Cline
- Cryptic species complex
- Encyclopedia of Life
- Endangered species
- Ring species
- Species naming
- Species problem
- Systematics

Notes and references

[1] Sahney, S., Benton, M.J. and Ferry, P.A. (2010). "Links between global taxonomic diversity, ecological diversity and the expansion of vertebrates on land" (http://rsbl.royalsocietypublishing.org/content/6/4/544.full.pdf+html) (PDF). *Biology Letters* **6** (4): 544–547. doi:10.1098/rsbl.2009.1024. PMC 2936204. PMID 20106856. .

[2] Sahney, S. and Benton, M.J. (2008). "Recovery from the most profound mass extinction of all time" (http://journals.royalsociety.org/content/qq5un1810k7605h5/fulltext.pdf) (PDF). *Proceedings of the Royal Society: Biological* **275** (1636): 759. doi:10.1098/rspb.2007.1370. PMC 2596898. PMID 18198148. .

[3] Goldenberg, Suzanne (2011-08-23). "Planet Earth is home to 8.7 million species, scientists estimate" (http://www.guardian.co.uk/environment/2011/aug/23/species-earth-estimate-scientists). *The Guardian* (London). . Retrieved 2011-08-23

[4] "Just How Many Species Are There, Anyway?" (http://www.sciencedaily.com/releases/2003/05/030526103731.htm). 2003-05-26. . Retrieved 2008-01-15

[5] Koch, H. 2010. Combining morphology and DNA barcoding resolves the taxonomy of Western Malagasy *Liotrigona* Moure, 1961. *African Invertebrates* **51** (2): 413-421. (http://www.africaninvertebrates.org.za/Koch_2010_51_2_474.aspx) PDF fulltext (http://www.tb1.ethz.ch/PublicationsEO/PDFpapers/Koch_AFRICAN_INVERTEBRATES_2010_51_413-421.pdf)

[6] De Queiroz K (December 2007). "Species concepts and species delimitation". *Syst. Biol.* **56** (6): 879–86. doi:10.1080/10635150701701083. PMID 18027281.

[7] Fraser C, Alm EJ, Polz MF, Spratt BG, Hanage WP (February 2009). "The bacterial species challenge: making sense of genetic and ecological diversity". *Science* **323** (5915): 741–6. doi:10.1126/science.1159388. PMID 19197054.

[8] de Queiroz K (May 2005). "Ernst Mayr and the modern concept of species" (http://www.pnas.org/cgi/pmidlookup?view=long&pmid=15851674). *Proc. Natl. Acad. Sci. U.S.A.* **102 Suppl 1**: 6600–7. doi:10.1073/pnas.0502030102. PMC 1131873. PMID 15851674. .

[9] http://www.ncbi.nlm.nih.gov/Taxonomy

[10] http://www.genome.jp/kegg/catalog/org_list.html

[11] http://www.uniprot.org/help/taxonomy

[12] Wilkins, John (2010-10-20). "How many species concepts are there?" (http://www.guardian.co.uk/science/punctuated-equilibrium/
     2010/oct/20/3). London: *The Guardian*. . Retrieved 2010-10-19.

[13] David I. Williamson (2003). *The Origins of Larvae*. Kluwer. ISBN 1-4020-1514-3.

[14] Darwin 1859 p.59 (http://darwin-online.org.uk/content/frameset?viewtype=side&itemID=F373&pageseq=59)

[15] Darwin 1871 p. 24 (http://darwin-online.org.uk/content/frameset?viewtype=text&itemID=F937.1&keywords=definition+species+
     of&pageseq=241)

[16] Louis Menand (2001) *The Metaphysical Club* New York: Farrar, Straus and Giroux 123–124

[17] Stackebrandt E, Goebel BM (1994). "Taxonomic note: a place for DNA-DNA reassociation and 16S rRNA sequence analysis in the present
     species definition in bacteriology". *Int. J. Syst. Bacteriol.* **44**: 846–9. doi:10.1099/00207713-44-4-846.

[18] Stackebrandt E, Ebers J (2006). "Taxonomic parameters revisited: tarnished gold standards". *Microbiol. Today* **33**: 152–5.

[19] Michael Ruse (August 1969). "Definitions of Species in Biology". *The British Journal for the Philosophy of Science* (Oxford University
     Press) **20** (2): 97–119. doi:10.1093/bjps/20.2.97. JSTOR 686173.

[20] Paraspecies

[21] James S. Albert; Roberto E. Reis. *Historical Biogeography of Neotropical Freshwater Fishes* (http://books.google.com/
     books?id=V8kZeHxkv9oC&pg=PA316&lpg=PA316&dq=evolutionary+species+concept+Alberty+Reis&source=bl&
     ots=Qd1DbXoGJv&sig=FausTKz_R9mhDEzHMpbk9Tq61K0&hl=en&ei=BipcTu-gIsq2twfM9PigDA&sa=X&oi=book_result&
     ct=result&resnum=1&ved=0CCAQ6AEwAA#v=onepage&q&f=false). .

[22] Ereshefsky M (2002). "Linnean ranks: vestiges of a bygone era". *Philosophy of Science* **69**: S305–S315. JSTOR 10.

[23] Current Results– Number of Species on Earth (http://www.currentresults.com/Environment-Facts/Plants-Animals/number-species.php)

[24] Sogin ML, Morrison HG, Huber JA, *et al.* (August 2006). "Microbial diversity in the deep sea and the underexplored "rare biosphere""
     (http://www.pnas.org/cgi/pmidlookup?view=long&pmid=16880384). *Proc. Natl. Acad. Sci. U.S.A.* **103** (32): 12115–20.
     doi:10.1073/pnas.0605127103. PMC 1524930. PMID 16880384. . Cheung L (Monday, 31 July 200). "Thousands of microbes in one gulp"
     (http://news.bbc.co.uk/1/hi/sci/tech/5232928.stm). BBC. .

[25] David L. Hawksworth (2001). "The magnitude of fungal diversity: the 1•5 million species estimate revisited" (http://journals.cambridge.
     org/action/displayAbstract?fromPage=online&aid=95069). *Mycological Research* **105** (12): 1422–1432. doi:10.1017/S0953756201004725.

[26] Acari at University of Michigan Museum of Zoology Web Page (http://insects.ummz.lsa.umich.edu/ACARI/index.html)

[27] Encyclopedia Smithsonian: Numbers of Insects (http://www.si.edu/Encyclopedia_SI/nmnh/buginfo/bugnos.htm)

[28] "Number of Species on Earth" (http://www.currentresults.com/Environment-Facts/Plants-Animals/number-species.php). Current
     Results. 2007-01-01. . Retrieved 2010-04-23.

[29] "Census of marine life" (http://www.coml.org/). Coml.org. . Retrieved 2010-04-23.

[30] Robin McKie and Zoe Corbyn (2005-09-25). "Discovery of new species and extermination at high rate" (http://www.guardian.co.uk/
     science/2005/sep/25/taxonomy.conservationandendangeredspecies). London: Guardian. . Retrieved 2010-04-23.

External links

- Stanford Encyclopedia of Philosophy entry: *Species* (http://plato.stanford.edu/entries/species/)
- Barcoding of species (http://www.barcodinglife.org/)
- European Species Names in Linnaean, Czech, English, German and French (http://www.finitesite.com/
  dandelion/Linnaeus.HTML)
- Catalogue of Life (http://www.catalogueoflife.org/)
- VisualTaxa (http://visualtaxa.redgolpe.com/)
- Speciation (http://users.rcn.com/jkimball.ma.ultranet/BiologyPages/S/Speciation.html)

Articles online

- Other Species Concepts (http://evolution.berkeley.edu/evosite/evo101/VA2OtherSpeciesConcept.shtml) -
  U.C. Berkeley
- "Gone" (http://www.motherjones.com/news/feature/2007/05/gone.html), *Mother Jones*, May/June 2007.
- 2003-12-31, ScienceDaily: Working On The 'Porsche Of Its Time': New Model For Species Determination
  Offered (http://www.sciencedaily.com/releases/2003/12/031231082553.htm)
- 2003-08-08, ScienceDaily: Cross-species Mating May Be Evolutionarily Important And Lead To Rapid Change
  (http://www.sciencedaily.com/releases/2003/08/030808081854.htm)
- 2004-01-09 ScienceDaily: Mayo Researchers Observe Genetic Fusion Of Human, Animal Cells; May Help
  Explain Origin Of AIDS (http://www.sciencedaily.com/releases/2004/01/040109064407.htm)

- 2000-09-18, ScienceDaily: Scientists Unravel Ancient Evolutionary History Of Photosynthesis (http://www. sciencedaily.com/releases/2000/09/000913211733.htm)

rue:Вид (біологія)

Cancellariidae

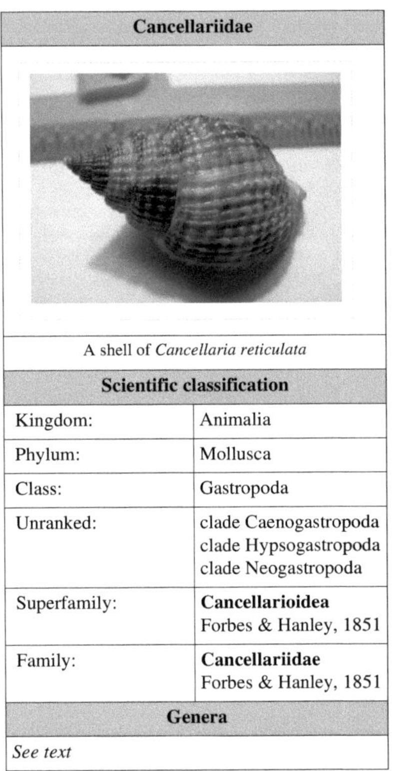

Cancellariidae	
A shell of *Cancellaria reticulata*	
Scientific classification	
Kingdom:	Animalia
Phylum:	Mollusca
Class:	Gastropoda
Unranked:	clade Caenogastropoda clade Hypsogastropoda clade Neogastropoda
Superfamily:	**Cancellarioidea** Forbes & Hanley, 1851
Family:	**Cancellariidae** Forbes & Hanley, 1851
Genera	
See text	

Cancellariidae, common name the **nutmeg snails** or **nutmeg shells**, are a family of small to medium-large sea snails, marine gastropod mollusks in the clade Neogastropoda. Their shells resemble a nutmeg seed.

Cancellariidae is the only family in the superfamily **Cancellarioidea**.

Distribution

This family occurs worldwide. Many species are found in deep water.

Taxonomy

This family consists of three following subfamilies (according to the taxonomy of the Gastropoda by Bouchet & Rocroi, 2005):

- Cancellariinae Forbes & Hanley, 1851 - synonym: Trigonostomatinae Cossmann, 1899
- Admetinae Troschel, 1865 - synonym: Paladmetidae Stephenson, 1941
- Plesiotritoninae Beu & Maxwell, 1937

Genera

Genera in the family Cancellariidae include: [1]

- *Admete* Krøyer, 1842
- *Admetula* Cossmann, 1889
- *Africotriton* Beu & Marshall, 1987
- *Agatrix* R. Petit, 1967
- *Antizafra* Finlay, 1926
- *Aphera* H. Adams & A. Adams, 1854
- *Arizelostoma* Iredale, 1936
- *Axelella* Petit, 1988
- *Bivetia* Jousseaume, 1887
- *Bivetiella* Wenz, 1943
- *Bivetopsia* Jousseaume, 1887
- *Bonellitia* Jousseaume, 1887
- *Brocchinia* Jousseaume, 1887
- *Cancellaphera* Iredale, 1930
- *Cancellaria* Lamarck, 1799
- *Cancellicula* Tabanelli, 2008
- *Crawfordina* Dall, 1919
- *Dellina* Beu, 1970
- *Euclia* H. Adams & A. Adams, 1854
- *Fusiaphera* Habe, 1961
- *Gerdiella* Olsson and Bayer, 1972
- *Gergovia* Cossmann, 1899
- *Habesolatia* Kuroda, 1965
- *Hertleinia* Marks, 1949
- *Inglisella* Finlay, 1924
- *Iphinoella* Habe, 1958 : synonym of *Iphinopsis*
- *Iphinopsis* Dall, 1924
- *Loxotaphrus* Harris, 1897
- *Massyla* H. Adams & A. Adams, 1854
- *Merica* H. Adams & A. Adams, 1854
- *Mericella* Thiele, 1929
- *Microcancilla* Dall, 1924
- *Microsveltia* Iredale, 1925

- *Mirandaphera* Bouchet & Petit, 2002
- *Momoebora* Kuroda & Habe, 1971
- *Narona* H. Adams & A. Adams, 1854
- *Neadmete* Habe, 1961
- *Nevia* Jousseaume, 1887
- *Nipponaphera* Habe, 1961
- *Nothoadmete* Oliver, 1982
- *Paladmete* Gardner, 1916
- *Pallidonia* Laseron, 1955
- *Pepta* Iredale, 1925
- *Perplicaria* Dall, 1890
- *Pisanella* Koenen, 1865
- *Plesiotriton* P. Fischer, 1884
- *Progabbia* Dall, 1918
- *Pseudobabylonella* Brunetti, Della Belle, Forli & Vecchi, 2009
- *Pyruclia* Olsson, 1932
- *Scalptia* Jousseaume, 1887
- *Solatia* Jousseaume, 1887
- *Solutosveltia* Habe, 1961
- *Sveltella* Cossmann, 1889
- *Sveltia* Jousseaume, 1887
- *Sydaphera* Iredale, 1929
- *Tribia* Jousseaume, 1887
- *Trigona* Perry, 1811
- *Trigonaphera* Iredale, 1936
- *Trigonostoma* de Blainville, 1825
- *Tritonium* Fabricius, 1780
- *Tritonoharpa* Dall, 1908
- *Ventrilia* Jousseaume, 1887
- *Vercomaris* Garrard, 1975
- *Waipaoa* Marwick, 1931
- *Zeadmete* Finlay, 1926

The following genus was also accepted in 1936 by the Royal Society of New Zealand [2]

- *Anapepta* Finlay, 1930 [3] [4]

References

[1] WoRMS : Cancellariidae; accessed 26 August 2010 (http://www.marinespecies.org/aphia.php?p=taxdetails&id=13600)

[2] Transactions and Proceedings of the Royal Society of New Zealand 1868-1961, Volume 65, 1936 (http://rsnz.natlib.govt.nz/volume/rsnz_65/rsnz_65_00_000870.html)

[3] "Cancellariidae" (http://www.itis.gov/servlet/SingleRpt/SingleRpt?search_topic=TSN&search_value=74336). Integrated Taxonomic Information System. .

[4] Powell A. W. B., *New Zealand Mollusca*, William Collins Publishers Ltd, Auckland, New Zealand 1979 ISBN 0-00-216906-1

- Verhecken A. (2007). Revision of the Cancellariidae (Mollusca, Neogastropoda, Cancellarioidea) of the eastern Atlantic (40°N-40°S) and the Mediterranean. Zoosystema : 29(2): 281-364 (http://www.mnhn.fr/museum/front/medias/publication/10531_z07n2a2.pdf)
- Hemmen J. (2007). Recent Cancellariidae. Wiesbaden, 428pp

Family_(biology)

In biological classification, **family** (Latin: *familia*) is

- a taxonomic rank. Other well-known ranks are life, domain, kingdom, phylum, class, order, genus, and species, with family fitting between order and genus. As for the other well-known ranks, there is the option of an immediately lower rank, indicated by the prefix *sub-*: subfamily (Latin: *subfamilia*).

- a taxonomic unit, a taxon, in that rank. In that case the plural is families (Latin *familiae*)

> *Example*: Walnuts and hickories belong to Juglandaceae, the walnut family.

What does and does not belong to each family is determined by a taxonomist. Similarly for the question if a particular family should be recognized at all. Often there is no exact agreement, with different taxonomists each taking a different position. There are no hard rules that a taxonomist needs to follow in describing or recognizing a family. Some taxa are accepted almost universally, while others are recognised only rarely.

History

The taxonomic term *familia* was first used by French botanist Pierre Magnol in his *Prodromus historiae generalis plantarum, in quo familiae plantarum per tabulas disponuntur* (1689) where he called the seventy-six groups of plants he recognised in his tables families (*familiae*). The concept of rank at that time was not yet settled, and in the preface to the *Prodromus* Magnol spoke of uniting his families into larger *genera*, which is far from how the term is used today.

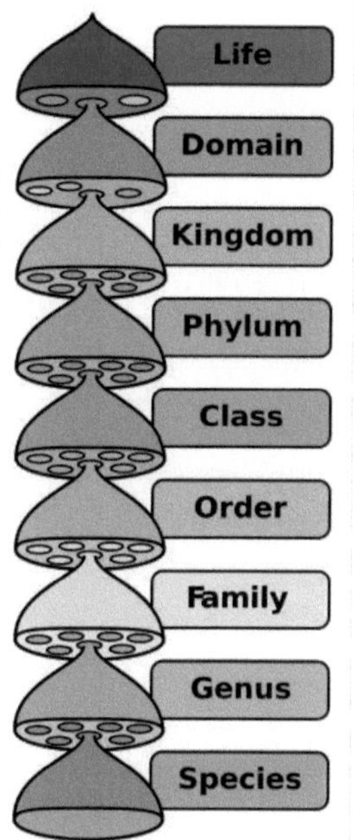

The hierarchy of biological classification's eight major taxonomic ranks, which is an example of definition by genus and differentia. An order contains one or more families. Intermediate minor rankings are not shown.

Carolus Linnaeus used the word *familia* in his *Philosophia botanica* (1751) to denote major groups of plants: trees, herbs, ferns, palms, and so on. He used this term only in the morphological section of the book, discussing the vegetative and generative organs of plants. Subsequently, in French botanical publications, from Michel Adanson's *Familles naturelles des plantes* (1763) and until the end of the 19th century, the word *famille* was used as a French equivalent of the Latin *ordo* (or *ordo naturalis*). In nineteenth century works such as the *Prodromus* of Augustin Pyramus de Candolle and the *Genera Plantarum* of George Bentham and Joseph Dalton Hooker this word *ordo* was used for what now is given the rank of family.

In zoology, the family as a rank intermediate between order and genus was introduced by Pierre André Latreille in his *Précis des caractères génériques des insectes, disposés dans un ordre naturel* (1796). He used families (some of them not named) in some but not in all his orders of "insects" (which then included all arthropods).

Uses

Families can be used for evolutionary, palaeontological and generic studies because they are more stable than lower taxonomic levels such as genera and species.[1] [2]

See also

- Systematics, the study of the diversity of life
- Cladistics, the classification of organisms by their order of branching in an evolutionary tree
- Phylogenetics, the study of evolutionary relatedness among various groups of organisms
- Taxonomy
- Virus classification
- List of Anuran families
- List of Testudines families
- List of fish families
- List of families of spiders

Compare:

- family
- protein family
- gene family

References

[1] Sarda Sahney, Michael J. Benton & Paul A. Ferry (2010). "Links between global taxonomic diversity, ecological diversity and the expansion of vertebrates on land" (http://rsbl.royalsocietypublishing.org/content/6/4/544.full.pdf+html) (PDF). *Biology Letters* **6** (4): 544–547. doi:10.1098/rsbl.2009.1024. PMC 2936204. PMID 20106856. .

[2] Sarda Sahney & Michael J. Benton (2008). "Recovery from the most profound mass extinction of all time" (http://journals.royalsociety.org/content/qq5un1810k7605h5/fulltext.pdf) (PDF). *Proceedings of the Royal Society B: Biological Sciences* **275** (1636): 759–765. doi:10.1098/rspb.2007.1370. PMC 2596898. PMID 18198148. .

rue:Родина (біологія)

Gastropoda

The **Gastropoda** or **gastropods**, more commonly known as **snails and slugs**, are a large taxonomic class within the phylum Mollusca. The class Gastropoda includes snails and slugs of all kinds and all sizes from microscopic to quite large. There are huge numbers of sea snails and sea slugs, as well as freshwater snails and freshwater limpets, and land snails and land slugs.

The class Gastropoda contains a vast total of named species, second only to the insects in overall number. The fossil history of this class goes all the way back to the Late Cambrian. There are 611 families of gastropods, of which 202 families are extinct, being found only in the fossil record.[1]

Gastropoda (previously known as **univalves** and sometimes spelled Gasteropoda) are a major part of the phylum Mollusca and are the most highly diversified class in the phylum, with 60,000 to 80,000[1] [2] living snail and slug species. The anatomy, behavior, feeding and reproductive adaptations of gastropods vary significantly from one clade or group to another. Therefore, it is difficult to state many generalities for all gastropods.

The class Gastropoda has an extraordinary diversification of habitats. Representatives live in gardens, in woodland, in deserts, and on mountains; in small ditches, great rivers and lakes; in estuaries, mudflats, the rocky intertidal, the sandy subtidal, in the abyssal depths of the oceans including the hydrothermal vents, and numerous other ecological niches, including parasitic ones.

Although the name "snail" can be, and often is, applied to all the members of this class, commonly this word means only those species with an external shell large enough that the soft parts can withdraw completely into it. Those gastropods without a shell, and those with only a very reduced or internal shell, are usually known as slugs.

The marine shelled species of gastropod include edible species such as abalone, conches, periwinkles, whelks, and numerous other sea snails that produce seashells which are coiled in the adult stage, even though in some cases the coiling may not be very visible, for example in cowries. There are also a number of families of species such as all the various limpets, where the shell is coiled only in the larval stage, and is a simple conical structure after that.

Etymology

The word "gastropod" is derived from the Ancient Greek words γαστήρ *(gastér,* stem: *gastr-)* "stomach", and πούς *(poús,* stem: *pod-)* "foot", hence stomach-foot. This is an anthropomorphic name, based on the fact that to humans it appears as if snails and slugs crawl on their bellies. In reality, snails and slugs have their stomach, the rest of their digestive system and all the rest of their viscera in a hump on the opposite, dorsal side of the body. In most gastropods this visceral hump is covered by, and contained within, the shell.

In the scientific literature, gastropods were described under the vernacular (French) name "gasteropodes" by Georges Cuvier in 1795.[3] The name was later Latinized.

The earlier name *univalve* means "one valve" or shell, in contrast to *bivalve* applied to mollusks such as clams and meaning that those animals possess two valves or shells.

Diversity

At all taxonomic levels, gastropods are second only to the insects in terms of their diversity.[4]

Gastropods are the class of molluscs which have the greatest numbers of named species. However estimates of the total number of gastropod species varies widely, depending on the cited sources. The number of gastropod species can be ascertained from estimates of the number of described species of Mollusca with accepted names: about 85,000[5] (minimum 50,000,[5] maximum 120,000[5]). But an estimate of the total number of Mollusca, including undescribed species, is about 240,000 species.[6] The estimate of 85,000 molluscs includes 24,000 described species of terrestrial gastropods.[5]

Different estimates for aquatic gastropods (based on different sources) give about 30,000 species[7] of marine gastropods, and about 5,000 species of freshwater and brackish gastropods.[7] The total number of recent species of freshwater snails is about 4,000.[8]

There are 444 recently extinct species of gastropods (extinct since the year 1500),[9] 18 species that are now extinct in the wild (but still existing in captivity)[9] and 69 "possibly extinct" species.[9]

The number of prehistoric (fossil) species of gastropods is at least 15,000 species.[10]

Habitat

Some of the more familiar and better-known gastropods are terrestrial gastropods (the land snails and slugs) and some live in freshwater, but more than two thirds of all named species live in a marine environment.

Gastropods have a worldwide distribution from the near Arctic and Antarctic zones to the tropics. They have become adapted to almost every kind of existence on earth, having colonized every medium available except the air.

In habitats where there is not enough calcium carbonate to build a really solid shell, such as on some acidic soils on land, there are still various species of slugs, and also some snails with a thin translucent shell, mostly or entirely composed of the protein conchiolin.

Snails such as *Sphincterochila boissieri* and *Xerocrassa seetzeni* have adapted to desert conditions, other snails have adapted to an existence in ditches, near deepwater hydrothermal vents, the pounding surf of rocky shores, caves, and many other diverse areas.

Anatomy

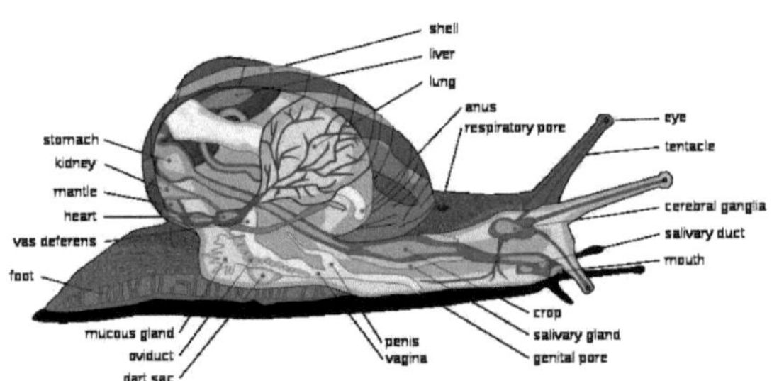

The anatomy of a common air-breathing land snail such as *Helix aspersa*. Note that much of this anatomy does not apply to gastropods in other clades or groups.

Snails are distinguished by an anatomical process known as torsion, where the visceral mass of the animal rotates 180° to one side during development, such that the anus is situated more or less above the head. This process is unrelated to the coiling of the shell, which is a separate phenomenon. Torsion is present in all gastropods, but the opisthobranch gastropods are secondarily de-torted to various degrees.[11] [12]

Torsion occurs in two mechanistic stages. The first is muscular and the second is mutagenetic. The effects of torsion are primarily physiological - the organism develops an asymmetrical nature with the majority of growth occurring on the left side. This leads to the loss of right-paired appendages (e.g. ctenidia (comb-like respiratory apparatus), gonads, nephridia, etc.). Furthermore, the anus becomes redirected to the same space as the head. This is speculated to have some evolutionary function, as prior to torsion, when retracting into the shell, first the posterior end would get pulled in, and then the anterior. Now, the front can be retracted more easily, perhaps suggesting a defensive purpose.

However, this "rotation hypothesis" is being challenged by the "asymmetry hypothesis" in which the gastropod mantle cavity originated from one side only of a bilateral set of mantle cavities.[13]

Gastropods typically have a well-defined head with two or four sensory tentacles with eyes, and a ventral foot, which gives them their name (Greek *gaster*, stomach, and *poda*, feet). The foremost division of the foot is called the **propodium**. Its function is to push away sediment as the snail crawls. The larval shell of a gastropod is called a protoconch.

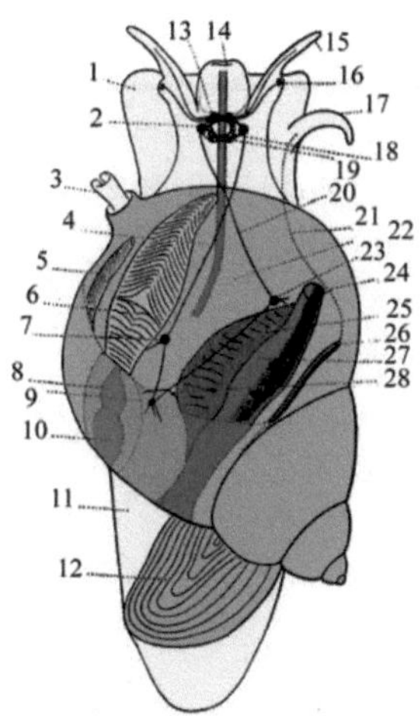

The anatomy of an aquatic snail with a gill, a male prosobranch gastropod. Note that much of this anatomy does not apply to gastropods in other clades.Light yellow - bodyBrown - shell and operculumGreen - digestive systemLight purple - gillsYellow - osphradiumRed - heartPink - Dark violet - 1. foot 2. cerebral ganglion 3. pneumostome 4. upper commissura 5. osphradium 6. gills 7. pleural ganglion 8. atrium of heart 9. visceral ganglion 10. ventricle 11. foot 12. operculum 13. brain 14. mouth 15. tentacle (chemosensory, 2 or 4) 16. eye 17. penis (everted, normally internal) 18. esophageal nerve ring 19. pedal ganglion 20. lower commissura 21. vas deferens 22. pallial cavity / mantle cavity / respiratory cavity 23. parietal ganglion 24. anus 25. hepatopancreas 26. gonad 27. rectum 28. nephridium

The shell

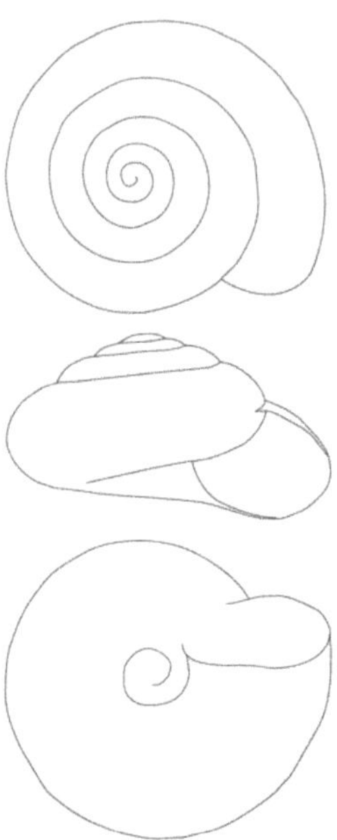

Most shelled gastropods have a one piece shell, typically coiled or spiraled. This coiled shell usually opens on the right-hand side (as viewed with the shell apex pointing upward). Numerous species have an operculum, which in many species acts as a trapdoor to close the shell. This is usually made of a horn-like material, but in some molluscs it is calcareous. In the land slugs, the shell is reduced or absent, and the body is streamlined.

Body wall

Some sea slugs are very brightly colored. This serves either as a warning, when they are poisonous or contain stinging cells, or to camouflage them on the brightly-colored hydroids, sponges and seaweeds on which many of the species are found.

Lateral outgrowths on the body of nudibranchs are called cerata. These contain a part of digestive gland, which is called the diverticula.

The shell of *Zonitoides nitidus*, a small land snail, has dextral coiling, which is typical (but not universal) in gastropod shells.
Upper image: dorsal view of the shell, showing the apex
Central image: lateral view showing the spire and aperture of the shell
Lower image: basal view showing the umbilicus

Sensory organs and nervous system

Sensory organs of gastropods include olfactory organs, eyes, statocysts and mechanoreceptors.[14] Gastropods have no hearing.[14]

In terrestrial gastropods (land snails and slugs), the olfactory organs, located on the tips of the 4 tentacles, are the most important sensory organ.[14] The chemosensory organs of opisthobranch marine gastropods are called rhinophores.

The majority of gastropods have simple visual organs, eye spots, that are situated either at the tip of the tentacles or the base of the tentacles. However "eyes" in gastropods range from these simple ocelli which cannot process an image being only able to distinguish light and dark, to more complex pit eyes, and even to lens eyes.[15] In land snails and slugs, vision is not the most important sense, because they are mainly nocturnal animals.[14]

The upper pair of tentacles on the head of *Helix pomatia* have eye spots, but the main sensory organs of the snail are sensory receptors for olfaction, situated in the epithelium of the tentacles.

The nervous system of gastropods includes the peripheral nervous system and the central nervous system. The central nervous system consist of ganglia connected by nerve cells. It includes paired ganglia: the cerebral ganglia, pedal ganglia, osphradial ganglia, pleural ganglia, parietal ganglia and the visceral ganglia. There are sometimes also buccal ganglia.[14]

Digestive system

The radula of a gastropod is usually adapted to the food that a species eats. The simplest gastropods are the limpets and abalones, herbivores that use their hard radula to rasp at seaweeds on rocks.

Many marine gastropods are burrowers, and have a siphon that extends out from the mantle edge. Sometimes the shell has a siphonal canal to accommodate this structure. A siphon enables the animal to draw water into their mantle cavity and over the gill. They use the siphon primarily to "taste" the water to detect prey from a distance. Gastropods with siphons tend to be either predators or scavengers.

Respiratory system

Almost all marine gastropods breathe with a gill, but many freshwater species, and the majority of terrestrial species, have a pallial lung. Gastropods with a lung belong to one group with common descent, the Pulmonata, however, gastropods with gills are paraphyletic. The respiratory protein in almost all gastropods is hemocyanin, but a pulmonate family Planorbidae have hemoglobin as respiratory protein.

In one large group of sea slugs, the gills are arranged as a rosette of feathery plumes on their backs, which gives rise to their other name, nudibranchs. Some nudibranchs have smooth or warty backs and have no visible gill mechanism, such that respiration may likely take place directly through the skin.

Circulatory system

Gastropods have open circulatory system and the transport fluid is hemolymph. Hemocyanin is present in the hemolymph as the respiratory pigment.

Excretory system

The primary organs of excretion in gastropods are nephridia, which produce either ammonia or uric acid as a waste product. The nephridium also plays an important role in maintaining water balance in freshwater and terrestrial species. Additional organs of excretion, at least in some species, include pericardial glands in the body cavity, and digestive glands opening into the stomach.

Reproductive system

Courtship is a part of mating behavior in some gastropods including some of the Helicidae. Again, in some land snails, an unusual feature of the reproductive system of gastropods is the presence and utilization of love darts.

In many marine gastropods other than the opisthobranchs, there are separate sexes; most land gastropods however are hermaphrodites.

Life cycle

The main aspects of the life cycle of gastropods include:

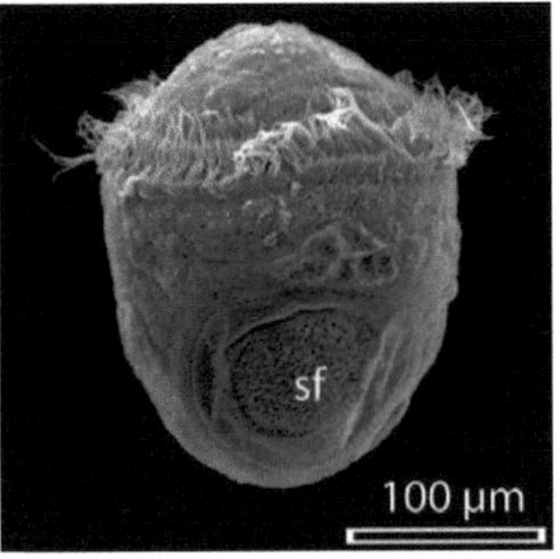

A 9-hour-old trochophore of *Haliotis asinina*
sf - shell field

- Egg laying and the eggs of gastropods
- The Embryonic development of gastropods
- The larvae or larval stadium: some gastropods may be trochophore and/or veliger
- Estivation and hibernation (each of these are present in some gastropods only)
- The growth of gastropods
- Courtship of gastropods and mating of gastropods: fertilisation is internal or external according to the species. External fertilisation is common in marine gastropods.

Feeding behavior

Marine gastropods include some that are herbivores, detritus feeders, predatory carnivores, scavengers, parasites, and also a few ciliary feeders, in which the radula is reduced or absent. Land-dwelling species can chew up leaves, bark, fruit and decomposing animals while marine species can scrape algae off the rocks on the sea floor. In some species that have evolved into endoparasites, such as *Parenteroxenos doglieli*, many of the standard gastropod features are strongly reduced or absent.

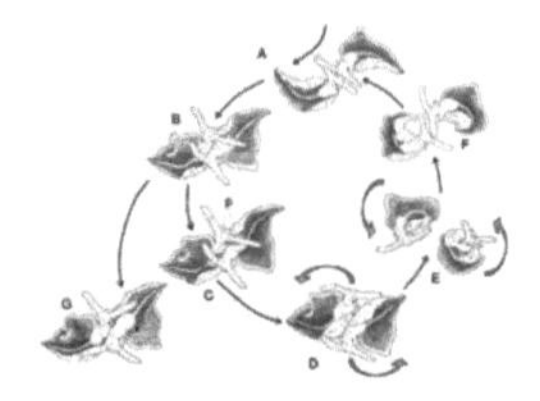

mating behaviour of *Elysia timida*

A few sea slugs are herbivores and some are carnivores. Many have distinct dietary preferences and regularly occur in close association with their food species.

Some predatory carnivorous gastropods include, for example: Cone shells, *Testacella*, *Daudebardia*, Ghost slug and others.

Genetics

Gastropods exhibit an important degree of variation in mitochondrial gene organization when compared to other animals.[16] Main events of gene rearrangement occurred at the origin of Patellogastropoda and Heterobranchia, whereas fewer changes occurred between the ancestors of Vetigastropoda (only tRNAs D, C and N) and Caenogastropoda (a large single inversion, and translocations of the tRNAs D and N).[16] Within Heterobranchia, gene order seems to be relatively conserved and gene rearrangements are mostly related with transposition of tRNA genes.[16]

Geological history

The first gastropods were exclusively marine, with the earliest representatives of the group appearing in the Late Cambrian (*Chippewaella*, *Strepsodiscus*). Early Cambrian forms like *Helcionella* and *Scenella* are no longer considered gastropods, and the tiny coiled *Aldanella* of earliest Cambrian time is probably not even a mollusk. By the Ordovician period the gastropods were a varied group present in a range of aquatic habitats. Commonly, fossil gastropods from the rocks of the early Palaeozoic era are too poorly preserved for accurate identification. Still, the Silurian genus *Poleumita* contains fifteen identified species. Fossil gastropods were less common during the Palaeozoic era than bivalves.

Fossil gastropod and attached mytilid bivalves on a Jurassic limestone bedding plane of the Matmor Formation in southern Israel.

Most of the gastropods of the Palaeozoic era belong to primitive groups, a few of which still survive today. By the Carboniferous period many of the shapes we see in living gastropods can be matched in the fossil record, but despite these similarities in appearance the majority of these older forms are not directly related to living forms. It was during the Mesozoic era that the ancestors of many of the living gastropods evolved.

One of the earliest known terrestrial (land-dwelling) gastropods is *Maturipupa*, which is found in the Coal Measures of the Carboniferous period in Europe, but relatives of the modern land snails are rare before the Cretaceous period, when the familiar *Helix* first appeared.

Helix aspersa: a European pulmonate land snail that has been accidentally introduced in many countries throughout the world.

Cepaea nemoralis: another European pulmonate land snail, which has been introduced to many other countries

In rocks of the Mesozoic era, gastropods are slightly more common as fossils, their shells are often well preserved. Their fossils occur in ancient beds deposited in both freshwater and marine environments. The "Purbeck Marble" of the Jurassic period and the "Sussex Marble" of the early Cretaceous period, which both occur in southern England, are limestones containing the tightly packed remains of the pond snail *Viviparus*.

Rocks of the Cenozoic era yield very large numbers of gastropod fossils, many of these fossils being closely related to modern living forms. The diversity of the gastropods increased markedly at the beginning of this era, along with that of the bivalves.

Certain trail-like markings preserved in ancient sedimentary rocks are thought to have been made by gastropods crawling over the soft mud and sand. Although these trails are of debatable origin, some of them do resemble the trails made by living gastropods today.

Gastropod fossils may sometimes be confused with ammonites or other shelled cephalopods. An example of this is *Bellerophon* from the limestones of the Carboniferous period in Europe, the shell of which is planispirally coiled and can be mistaken for the shell of a cephalopod.

Gastropods are one of the groups that record the changes in fauna caused by the advance and retreat of the Ice Sheets during the Pleistocene epoch.

Taxonomy

Since Darwin, biological taxonomy has tried to reflect the presumed phylogeny of organisms, i.e. the tree of life. The classifications used in taxonomy attempt to represent the precise interrelatedness of the various species. The taxonomy of the Gastropoda as shown in various texts can differ in major ways.

In the older classification of the gastropods, there were four subclasses:[17]

A group of fossil shells of *Turritella cingulifera* from the Pliocene of Cyprus.

- Opisthobranchia (gills to the right and behind the heart).
- Gymnomorpha (no shell)
- Prosobranchia (gills in front of the heart).
- Pulmonata (with a lung instead of gills)

The taxonomy of the Gastropoda is under constant revision, and more and more of the old taxonomy is being abandoned, as the results of DNA studies slowly become clearer. Nevertheless a few of the older terms such as "opisthobranch" and "prosobranch" are still sometimes used in a descriptive way.

New insights based on DNA sequencing of gastropods have produced some revolutionary new taxonomic insights. In the case of the Gastropoda, the taxonomy is now gradually being rewritten to embody strictly monophyletic groups (only one lineage of gastropods in each group). Integrating new findings into a working taxonomy will continue to be a challenge in coming years. Consistent ranks within the taxonomy at the level of subclass, superorder, order and suborder have already been abandoned as unworkable. Ongoing revisions of the higher taxonomic levels are to be expected in the near future.

Convergent evolution, which appears to exist at especially high frequency in the class Gastropoda class, may account for the observed differences between the older phylogenies which were based on morphological data, and more recent gene-sequencing studies.

Five different views of a shell of a *Fulguropsis* species

Bouchet & Rocroi (2005)[1] [18] made sweeping changes in the systematics, resulting in a taxonomy that is a step closer to the evolutionary history of the phylum.

The Bouchet & Rocroi classification system is based partly on the older systems of classification, and partly on new cladistic research. In the past, the taxonomy of gastropods was largely based on phenetic morphological characters of the taxa. The recent advances are more based on molecular characters from DNA[19] and RNA research. This has made the taxonomical ranks and their hierarchy controversial. The debate about these issues is not likely to end soon.

In the Bouchet, Rocroi *et al.* taxonomy, the authors have used **unranked** clades for taxa above the rank of superfamily (replacing the ranks suborder, order, superorder and subclass), while using the traditional Linnaean approach for all taxa below the rank of superfamily. Whenever monophyly has not been tested, or is known to be paraphyletic or polyphyletic, the term "group" or "informal group" has been used. The classification of families into subfamilies is often not well resolved, and should be regarded as the best possible hypothesis.

In 2004, Brian Simison and David R. Lindberg showed possible diphyletic origins of the Gastropoda based on mitochondrial gene order and amino acid sequence analyses of complete genes.[20]

References

This article incorporates CC-BY-2.0 text from the reference.[16]

[1]   Bouchet P. & Rocroi J.-P. (Ed.); Frýda J., Hausdorf B., Ponder W., Valdes A. & Warén A. 2005. *Classification and nomenclator of gastropod families*. Malacologia: International Journal of Malacology, 47(1-2). ConchBooks: Hackenheim, Germany. ISBN 3-925919-72-4. 397 pp. vliz.be (http://www.vliz.be/Vmdcdata/imis2/ref.php?refid=78278)

[2]   Britannica online: abundance of the Gastropoda (http://www.britannica.com/EBchecked/topic/226777/gastropod/35708/ Distribution-and-abundance)

[3]   (French) Cuvier G. (1795). "Second mémoire sur l'organisation et les rapports des animaux à sang blanc, dans lequel on traite de la structure des Mollusques et de leur division en ordres, lu à la Société d'histoire naturelle de Paris, le 11 Prairial, an III". *Magazin Encyclopédique, ou Journal des Sciences, des Lettres et des Arts* 2: 433-449. page 448 (http://www.archive.org/stream/ magazinencyclop12pari#page/448/mode/1up).

[4]   McArthur, A.G.; M.G. Harasewych (2003). "Molecular systematics of the major lineages of the Gastropoda.". *Molecular Systematics and Phylogeography of Mollusks*. Washington: Smithsonian Books. pp. 140–160.

[5]   Chapman, A.D. (2009). Numbers of Living Species in Australia and the World, 2nd edition (http://www.environment.gov.au/biodiversity/ abrs/publications/other/species-numbers/2009/04-02-groups-invertebrates.html#mollusca). Australian Biological Resources Study, Canberra. Accessed 12 January 2010. ISBN 978 0 642 56860 1 (printed); ISBN 978 0 642 56861 8 (online).

[6]   Appeltans W., Bouchet P., Boxshall G.A., Fauchald K., Gordon D.P., Hoeksema B.W., Poore G.C.B., van Soest R.W.M., Stöhr S., Walter T.C., Costello M.J. (eds) (2011). World Register of Marine Species. Accessed at marinespecies.org (http://www.marinespecies.org) on 2011-03-07.

[7]   "gastropod" (http://www.britannica.com/EBchecked/topic/226777/gastropod). (2010). In Encyclopædia Britannica. Retrieved March 05, 2010, from Encyclopædia Britannica Online.

[8]   Strong E. E., Gargominy O., Ponder W. F. & Bouchet P. (2008). "Global Diversity of Gastropods (Gastropoda; Mollusca) in Freshwater". *Hydrobiologia* 595: 149-166. http://hdl.handle.net/10088/7390 doi:10.1007/s10750-007-9012-6.

[9] Régnier C., Fontaine B. & Bouchet P. (2009). "Not Knowing, Not Recording, Not Listing: Numerous Unnoticed Mollusk Extinctions". *Conservation Biology* **23**(5): 1214-1221. PubMed (http://www.ncbi.nlm.nih.gov/pubmed/19459894), doi:10.1111/j.1523-1739.2009.01245.x.

[10] (Spanish) Nájera J. M. (1996). "Moluscos del suelo como plagas agrícolas y cuarentenarias". *X Congreso Nacional Agronómico / II Congreso de Suelos 1996* 51-56. PDF (http://www.bio-nica.info/biblioteca/Monje1996.pdf)

[11] Kay, A.; Wells, F. E.; Poder, W. F. (1998). "Class Gastropoda". In Beesley, P. L.; Ross, G. J. B.; Wells, A.. *Mollusca: The Southern Synthesis. Fauna of Australia.* CSIRO Publishing. pp. 565–604. ISBN 0 643 05756 0.

[12] Brusca, R. C.; Brusca, G. J. (2003). "Phylum Mollusca". *Invertebrates.* Sinauer Associates, Inc.. pp. 701–769. ISBN 0-87893-097-3.

[13] Louise R. Page (2006). "Modern insights on gastropod development: Reevaluation of the evolution of a novel body plan" (http://intl-icb.oxfordjournals.org/cgi/content/full/46/2/134). *Integrative and Comparative Biology* **46** (2): 134–143. doi:10.1093/icb/icj018. .

[14] Chase R.: *Sensory Organs and the Nervous System.* in Barker G. M. (ed.): *The biology of terrestrial molluscs.* CABI Publishing, Oxon, UK, 2001, ISBN 0-85199-318-4. 1-146, cited pages: 179-211.

[15] Götting, Klaus-Jürgen (1994). "Schnecken". In Becker, U., Ganter, S., Just, C. & Sauermost, R.. *Lexikon der Biologie.* Heidelberg: Spektrum Akademischer Verlag. ISBN 3-86025-156-2.

[16] Cunha R. L., Grande C. & Zardoya R. (23 August 2009). "Neogastropod phylogenetic relationships based on entire mitochondrial genomes". *BMC Evolutionary Biology* 2009, **9**: 210. doi:10.1186/1471-2148-9-210

[17] Paul Jeffery.*Suprageneric classification of class Gastropoda.* The Natural History Museum, London, 2001.

[18] Poppe G.T. & Tagaro S.P. 2006. The new classification of Gastropods according to Bouchet & Rocroi, 2005. Visaya, février 2006: 10 pp. journal-malaco.fr (http://www.journal-malaco.fr/bouchet&rocroi_2005_Visaya.pdf)

[19] Elpidio A. Remigio and Paul D.N. Hebert (2003). "Testing the utility of partial COI sequences for phylogenetic (full text on line)" (http://www.bolinfonet.org/pdf/MPEVsnailpaper.pdf). *Molecular Phylogenetics and Evolution* **29** (3): 641–647. doi:10.1016/S1055-7903(03)00140-4. PMID 14615199. .

[20] Unitas malacologica, Newsletter number 21 december 2004 - a .pdf file (http://www.ucd.ie/cobid/unitas/newsletter/UMNewsletter_21.pdf)

- Abbott, R. T. (1989): *Compendium of Landshells. A color guide to more than 2,000 of the World's Terrestrial Shells.* 240 S., American Malacologists. Melbourne, Fl, Burlington, Ma. ISBN 0-915826-23-2
- Abbott, R. T. & Dance, S. P. (1998): *Compendium of Seashells. A full-color guide to more than 4,200 of the world's marine shells.* 413 S., Odyssey Publishing. El Cajon, Calif. ISBN 0-9661720-0-0
- Parkinson, B., Hemmen, J. & Groh, K. (1987): *Tropical Landshells of the World.* 279 S., Verlag Christa Hemmen. Wiesbaden. ISBN 3-925919-00-7
- Ponder, W. F. & Lindberg, D. R. (1997): *Towards a phylogeny of gastropod molluscs: an analysis using morphological characters. Zoological Journal of the Linnean Society,* **119** 83–265.
- Robin, A. (2008): *Encyclopedia of Marine Gastropods.* 480 S., Verlag ConchBooks. Hackenheim. ISBN 978-3-939767-09-1

External links

- Gastropod reproductive behavior (http://www.scholarpedia.org/article/Gastropod_reproductive_behavior)
- Reconstructions of fossil gastropods (http://www.emilydamstra.com/portfolio.php?subcat=14)
- 2004 Linnean taxonomy of gastropods (http://www.manandmollusc.net/advanced_introduction/gastropod_taxonomy_1.html)
- Webster S. J. & Fiorito G. (October 2001) "Socially guided behaviour in non-insect invertebrates". *Animal Cognition* **4**(2): 69–79. doi: 10.1007/s100710100108 - An article about social learning also in gastropods.

bjn:Gondang

Mollusca

The **Mollusca** (pronounced /məˈlʌskə/), common name **molluscs** or **mollusks**[1] (pronounced /ˈmɒləsks/), is a large phylum of invertebrate animals. There are around 85,000 recognized extant species of molluscs. Mollusca is the largest marine phylum, comprising about 23% of all the named marine organisms. Numerous molluscs also live in freshwater and terrestrial habitats. Molluscs are highly diverse, not only in size and in anatomical structure, but also in behaviour and in habitat. The phylum is typically divided into nine or ten taxonomic classes, of which two are entirely extinct. Cephalopod molluscs such as squid, cuttlefish and octopus are among the most neurologically advanced of all invertebrates – and either the giant squid or the colossal squid is the largest known invertebrate species. The gastropods (snails and slugs) are by far the most numerous molluscs in terms of classified species, and account for 80% of the total.

Molluscs have such a varied range of body structures that it is difficult to find defining characteristics that apply to all modern groups. The two most universal features are a mantle with a significant cavity used for breathing and excretion, and the structure of the nervous system. As a result of this wide diversity, many textbooks base their descriptions on a hypothetical "generalized mollusc". This has a single, "limpet-like" shell on top, which is made of proteins and chitin reinforced with calcium carbonate, and is secreted by a mantle that covers the whole upper surface. The underside of the animal consists of a single muscular "foot". Although molluscs are coelomates, the coelom is very small, and the main body cavity is a hemocoel through which blood circulates – molluscs' circulatory systems are mainly open. The "generalized" mollusc's feeding system consists of a rasping "tongue" called a radula and a complex digestive system in which exuded mucus and microscopic, muscle-powered "hairs" called cilia play various important roles. The "generalized mollusc" has two paired nerve cords, or three in bivalves. The brain, in species that have one, encircles the esophagus. Most molluscs have eyes, and all have sensors that detect chemicals, vibrations and touch. The simplest type of molluscan reproductive system relies on external fertilization, but there are more complex variations. All produce eggs, from which may emerge trochophore larvae, more complex veliger larvae, or miniature adults.

A striking feature of molluscs is the use of the same organ for multiple functions. For example: the heart and nephridia ("kidneys") are important parts of the reproductive system as well as the circulatory and excretory systems; in bivalves, the gills both "breathe" and produce a water current in the mantle cavity, which is important for excretion and reproduction.

There is good evidence for the appearance of gastropods, cephalopods and bivalves in the Cambrian period **unknown operator: u'cambrian'** to 488.3 [2] million years ago. However the evolutionary history both of molluscs' emergence from the ancestral Lophotrochozoa and of their diversification into the well-known living and fossil forms are still subjects of vigorous debate among scientists.

Molluscs have been and still are an important food source for anatomically modern humans. However there is a risk of food-poisoning from toxins that accumulate in molluscs under certain conditions, and many countries have regulations that aim to minimize this risk. Molluscs have for centuries also been the source of important luxury goods, notably pearls, mother of pearl, Tyrian purple dye, and sea silk. Their shells have also been used as a money in some pre-industrial societies, although shell "currencies" have severe limitations compared with government-backed money.

Mollusc species can also represent hazards or pests for human activities. The bite of the blue-ringed octopus is often fatal, and that of *Octopus apollyon* causes inflammation that can last for over a month. Stings from a few species of large tropical cone shells can also kill, but their sophisticated though easily produced venoms have become important tools in neurological research. Schistosomiasis (also known as bilharzia, bilharziosis or snail fever) is transmitted to humans via water snail hosts, and affects about 200 million people. Snails and slugs can also be serious agricultural pests, and accidental or deliberate introduction of some snail species into new environments has seriously damaged some ecosystems.

Etymology

The words *mollusc* and mollusk are both derived from the French *mollusque,* which originated from the Latin *molluscus,* from *mollis,* soft. *Molluscus* was itself an adaptation of Aristotle's τὰ μαλάκια, "the soft things", which he applied to cuttlefish.[3] The scientific study of molluscs is known as malacology.[4]

Definition

The two most universal features of the body structure of molluscs are a mantle with a significant cavity used for breathing and excretion, and the organization of the nervous system. The most abundant metallic element in molluscs is calcium.[5]

Molluscs have developed such a varied range of body structures that it is difficult to find synapomorphies (defining characteristics) that apply to all modern groups.[6] The most general characteristic of molluscs is that they are unsegmented and bilaterally symmetrical.[7] The following are present in all modern molluscs:[8] [9]

- The dorsal part of the body wall is a mantle (or pallium) which secretes calcareous spicules, plates or shells. It overlaps the body with enough spare room to form a mantle cavity.
- The anus and genitals open into the mantle cavity.
- There are two pairs of main nerve cords.[9]

Other characteristics that commonly appear in textbooks have significant exceptions:

				Class			
Characteristic[8]	Aplacophora[10]	Polyplacophora[11]	Monoplacophora[12]	Gastropoda[13]	Cephalopoda[14]	Bivalvia[15]	Scaphopoda[16]
Radula, a rasping "tongue" with chitinous teeth	Absent in 20% of Neomeniomorpha	Yes	Yes	Yes	Yes	No	Internal, cannot extend beyond body
Broad, muscular foot	Reduced or absent	Yes	Yes	Yes	Modified into arms	Yes	Small, only at "front" end
Dorsal concentration of internal organs (visceral mass)	Not obvious	Yes	Yes	Yes	Yes	Yes	Yes
Large digestive ceca	No ceca in some aplacophora	Yes	Yes	Yes	Yes	Yes	No
Large complex metanephridia ("kidneys")	None	Yes	Yes	Yes	Yes	Yes	Small, simple

Diversity

Estimates of accepted described living species of molluscs vary from 50,000 to a maximum of 120,000 species.[18] In 2009 Chapman estimated the number of described living species at 85,000.[18] Haszprunar in 2001 estimated about 93,000 named species,[19] which include 23% of all named marine organisms.[20] Molluscs are second only to arthropods in numbers of living animal species[17] —far behind the arthropods' 1,113,000 but well ahead of chordates' 52,000.[21] It has been estimated that there are about 200,000 living species in total,[18][22] and 70,000 fossil species,[8] although the total number of mollusc species that ever existed, whether or not preserved, must be many times greater than the number alive today.[23]

About 80% of all known mollusc species are gastropods (snails and slugs), including the cowry (a sea snail) pictured here.[17]

Molluscs have more varied forms than any other animal phylum. They include snails, slugs and other gastropods; clams and other bivalves; squids and other cephalopods; and other lesser-known but similarly distinctive sub-groups. The majority of species still live in the oceans, from the seashores to the abyssal zone, but some form a significant part of the freshwater fauna and the terrestrial ecosystems. Molluscs are extremely diverse in tropical and temperate regions but can be found at all latitudes.[6] About 80% of all known mollusc species are gastropods.[17] Cephalopoda such as squid, cuttlefish and octopus are among the neurologically most advanced of all invertebrates.[24] The giant squid, which until recently had not been observed alive in its adult form,[25] is one of the largest invertebrates. However a recently caught specimen of the colossal squid, 10 metres (33 ft) long and weighing 500 kilograms (1100 lb), may have overtaken it.[26]

Freshwater and terrestrial molluscs appear exceptionally vulnerable to extinction. Estimates of the numbers of non-marine molluscs vary widely, partly because many regions have not been thoroughly surveyed. There is also a shortage of specialists who can identify all the animals in any one area to species. However, in 2004 the IUCN Red List of Threatened Species included nearly 2,000 endangered non-marine molluscs. For comparison, the great majority of mollusc species are marine but only 41 of these appeared on the 2004 Red List. 42% of recorded extinctions since the year 1500 are of molluscs, almost entirely non-marine species.[27]

A "generalized mollusc"

Further information: Mollusc shell

1

2

3

4

5

6

7

8

9

10

11

12

13

14

15

16

17

18

Digestive & excretory system

Circulatory & respiratory

Central nervous system

Reproductive system

1 Radula

2 Mouth

3 Shell

4 Stomach

5 Gonad

6 Heart

7 Coelom

8 Nephridium

9 Mantle

10 Mantle cavity

11 Anus

12 Gill

13 Foot

14 Hemocoel

15 Pedal nerve cord

16 Gut

17 Visceral nerve cord

18 Nerve ring

A generalized mollusc[28]

Because of the great range of anatomical diversity among molluscs, many textbooks start the subject by describing a hypothetical "generalized mollusc" to illustrate the most common features found within the phylum. The depiction is rather similar to modern monoplacophorans, and some suggest it *may* resemble very early molluscs.[6] [9] [12] [29]

The generalized mollusc has a single, "limpet-like" shell on top. The shell is secreted by a mantle that covers the upper surface. The underside consists of a single muscular "foot".[9] The visceral mass, or visceropallium, is the soft, non-muscular metabolic region of the mollusc. It contains the body organs.[7]

Mantle and mantle cavity

The mantle cavity is a fold in the mantle that encloses a significant amount of space. It is lined with epidermis. It is exposed, according to habitat, to sea, fresh water or air. The cavity was at the rear in the earliest molluscs but its position now varies from group to group. The anus, a pair of osphradia (chemical sensors) in the incoming "lane", the hindmost pair of gills and the exit openings of the nephridia ("kidneys") and gonads (reproductive organs) are in the mantle cavity.[9] The whole soft body of bivalves lies within an enlarged mantle cavity.[7]

Shell

The mantle edge secretes a shell (secondarily absent in a number of taxonomic groups, such as the nudibranchs[7]) that consists of mainly chitin and conchiolin (a protein) hardened with calcium carbonate),[9] [30] except that the outermost layer in almost all cases is all conchiolin (see periostracum).[9] Molluscs never use phosphate to construct their hard parts,[31] with the questionable exception of *Cobcrephora*.[32] While most mollusc shells are composed mainly of aragonite, those gastropods that lay eggs with a hard shell use calcite (sometimes with traces of aragonite) to construct the eggshells.[33]

The shell consists of three layers : the outer layer (the periostracum) made of organic matter, a middle layer made of columnar calcite and an inner layer consisting of laminated calcite, that is often nacreous.[7]

Foot

The underside consists of a muscular foot, which has adapted to different purposes in different classes.[34] :4 The foot carries a pair of statocysts, which act as balance sensors. In gastropods, it secretes mucus as a lubricant to aid movement. In forms that have only a top shell, such as limpets, the foot acts as a sucker attaching the animal to a hard surface, and the vertical muscles clamp the shell down over it; in other molluscs, the vertical muscles pull the foot and other exposed soft parts into the shell.[9] In bivalves, the foot is adapted for burrowing into the sediment;[34] :4 in cephalopods it is used for jet propulsion,[34] :4 and the tentacles and arms are derived from the foot.[35]

Circulation

Molluscs' circulatory systems are mainly open. Although molluscs are coelomates, their coeloms are reduced to fairly small spaces enclosing the heart and gonads. The main body cavity is a hemocoel through which blood and coelomic fluid circulate and which encloses most of the other internal organs. These hemocoelic spaces act as an efficient hydrostatic skeleton.[7] The blood contains the respiratory pigment hemocyanin as an oxygen-carrier. The heart consists of one or more pairs of atria (auricles), which receive oxygenated blood from the gills and pump it to the ventricle, which pumps it into the aorta (main artery), which is fairly short and opens into the hemocoel.[9]

The atria of the heart also function as part of the excretory system by filtering waste products out of the blood and dumping it into the coleom as urine. A pair of nephridia ("little kidneys") to the rear of and connected to the coelom extracts any re-usable materials from the urine and dumps additional waste products into it, and then ejects it via tubes that discharge into the mantle cavity.[9]

Respiration

Most molluscs have only one pair of gills, or even only one gill. Generally the gills are rather like feathers in shape, although some species have gills with filaments on only one side. They divide the mantle cavity so that water enters near the bottom and exits near the top. Their filaments have three kinds of cilia, one of which drives the water current through the mantle cavity, while the other two help to keep the gills clean. If the osphradia detect noxious chemicals or possibly sediment entering the mantle cavity, the gills' cilia may stop beating until the unwelcome intrusions have ceased. Each gill has an incoming blood vessel connected to the hemocoel and an outgoing one to the heart.[9]

Eating, digestion, and excretion

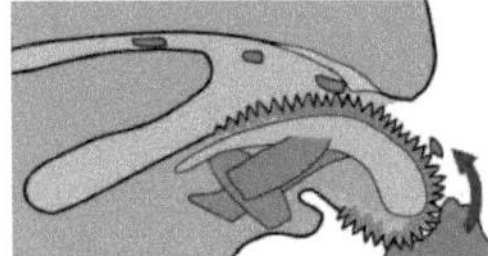

= Food

= Radula

= Odontophore "belt"

= Muscles

Snail radula at work.

Most molluscs have muscular mouths with radulae, "tongues" bearing many rows of chitinous teeth, which are replaced from the rear as they wear out. The radula primarily functions to scrape bacteria and algae off rocks. This radula is associated with the odontophore, a cartilaginous supporting organ[7]

Molluscs mouths also contain glands that secrete slimy mucus, to which the food sticks. Beating cilia (tiny "hairs") drive the mucus towards the stomach, so that the mucus forms a long string.[9]

At the tapered rear end of the stomach and projecting slightly into the hindgut is the prostyle, a backward-pointing cone of feces and mucus, which is rotated by further cilia so that it acts as a bobbin, winding the mucus string onto itself. Before the mucus string reaches the prostyle, the acidity of the stomach makes the mucus less sticky and frees particles from it.[9]

The particles are sorted by yet another group of cilia, which send the smaller particles, mainly minerals, to the prostyle so that eventually they are excreted, while the larger ones, mainly food, are sent to the stomach's cecum (a pouch with no other exit) to be digested. The sorting process is by no means perfect.[9]

Periodically, circular muscles at the hindgut's entrance pinch off and excrete a piece of the prostyle, preventing the prostyle from growing too large. The anus is in the part of the mantle cavity that is swept by the outgoing "lane" of the current created by the gills. Carnivorous molluscs usually have simpler digestive systems.[9]

As the head has largely disappeared in bivalves, their mouth has been equipped with labial palps (two on each side of the mouth) to collect the detritus from its mucus.[7]

Nervous system

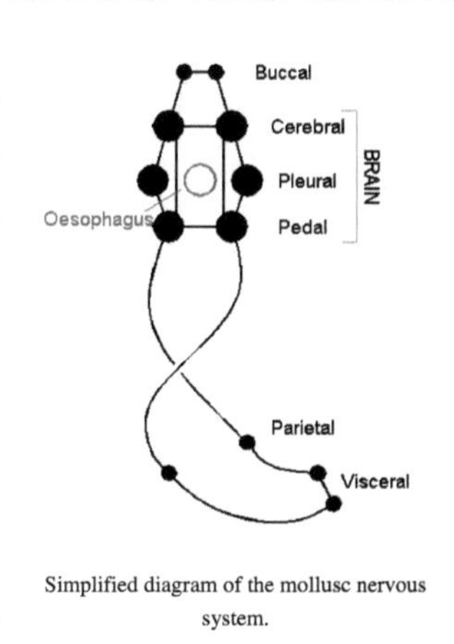

Simplified diagram of the mollusc nervous system.

Molluscs have two pairs of main nerve cords (three in bivalves) the visceral cords serving the internal organs and the pedal ones serving the foot. Both pairs run below the level of the gut, and include ganglia as local control centers in important parts of the body. Most pairs of corresponding ganglia on both sides of the body are linked by commissures (relatively large bundles of nerves). The only ganglia above the gut are the cerebral ganglia, which sit above the esophagus (gullet) and handle "messages" from and to the eyes. The pedal ganglia, which control the foot, are just below the esophagus and their commissure and connections to the cerebral ganglia encircle the esophagus in a nerve ring.[9]

The brain, in species that have one, encircles the esophagus. Most molluscs have a head with eyes, and all have a pair of sensor-containing tentacles, also on the head, that detect chemicals, vibrations and touch.[9]

Reproduction

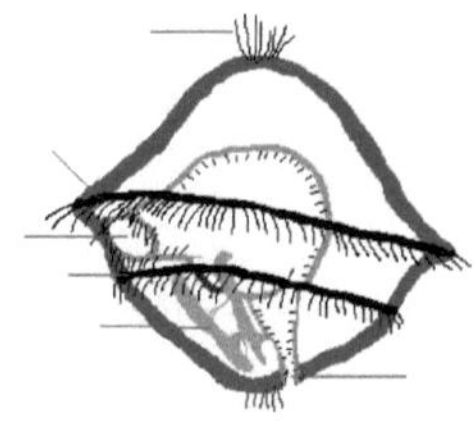

Apical tuft (cilia)

Prototroch (cilia)

Stomach

Mouth

Metatroch (cilia)

Mesoderm

Anus

/// = cilia

Trochophore larva[36]

The simplest molluscan reproductive system relies on external fertilization, but there are more complex variations. All produce eggs, from which may emerge trochophore larvae, more complex veliger larvae, or miniature adults. Two gonads sit next to the coelom, a small cavity that surrounds the heart and shed ova or sperm into the coloem, from which the nephridia extract them and emit them into the mantle cavity. Molluscs that use such a system remain of one sex all their lives and rely on external fertilization. Some molluscs use internal fertilization and/or are hermaphrodites, functioning as both sexes; both of these methods require more complex reproductive systems.[9]

The most basic molluscan larva is a trochophore, which is planktonic and feeds on floating food particles by using the two bands of cilia round its "equator" to sweep food into the mouth, which uses more cilia to drive them into the stomach, which uses further cilia to expel undigested remains through the anus. New tissue grows in the bands of mesoderm in the interior, so that the apical tuft and anus are pushed further apart as the animal grows. The trochophore stage is often succeeded by a veliger stage in which the prototroch, the "equatorial" band of cilia nearest the apical tuft, develops into the velum ("veil"), a pair of cilia-bearing lobes with which the larva swims. Eventually the larva sinks to the seafloor and metamorphoses into the adult form. Whilst metamorphosis is the usual state in molluscs, the cephalopods differ in exhibiting direct development: the hatchling is a 'miniaturized' form of the adult.[37]

Ecology

Feeding

Most molluscs are herbivorous, grazing on algae. Two feeding strategies are predominant: some feed on microscopic, filamentous algae, often using their radula as a 'rake' to comb up filaments from the sea floor. Others feed on macroscopic 'plants' such as kelp, rasping the plant itself with its radula. To employ this strategy, the plant has to be large enough for the mollusc to 'sit' on; therefore smaller macroscopic plants enjoy less molluscan herbivory than their larger counterparts.[38] Naturally, there are exceptions; the cephalopods are primarily (perhaps entirely) predatory, and the radula takes a secondary role to the jaws and tentacles in food acquisition. The monoplacophoran *Neopilina* uses its radula in the usual fashion, but its diet includes protists such as the xenophyophore *Stannophyllum*.[39] Sacoglossan nudibranchs suck the sap from algae, using their one-row radula to pierce the cell walls,[40] whereas dorid nudibranchs and some Vetigastrpods feed on sponges[41] [42] and others feed on hydroids.[43] (An extensive list of molluscs with unusual feeding habits is available in the appendix of GRAHAM, A. (1955). "Molluscan diets"[44]. *Journal of Molluscan Studies* **31** (3–4): 144..)

Classification

Opinions vary about the number of classes of molluscs—for example the table below shows eight living classes,[19] and two extinct ones. Although they are unlikely to form a clade, some older works combine the Caudofoveata and solenogasters into one class, the Aplacophora.[10] [29] Two of the commonly recognized "classes" are known only from fossils.[17]

Class	Major organisms	Described living species[19]	Distribution
Caudofoveata[10]	worm-like organisms	120	seabed 200–3000 metres (660–9800 ft)
Solenogastres[10]	worm-like organisms	200	seabed 200–3000 metres (660–9800 ft)
Polyplacophora[11]	chitons	1,000	rocky tidal zone and seabed
Monoplacophora[12]	An ancient lineage of molluscs with cap-like shells	31	seabed 1800–7000 metres (5900–23000 ft); one species 200 metres (660 ft)
Gastropoda[45]	All the snails and slugs including abalone, limpets, conch, nudibranchs, sea hares, sea butterfly	70,000	marine, freshwater, land
Cephalopoda[46]	squid, octopus, cuttlefish, nautilus	900	marine
Bivalvia[47]	clams, oysters, scallops, geoducks, mussels	20,000	marine, freshwater
Scaphopoda[16]	tusk shells	500	marine 6–7000 metres (20–23000 ft)
Rostroconchia †[48]	fossils; probable ancestors of bivalves	extinct	marine
Helcionelloida †[49]	fossils; snail-like organisms such as *Latouchella*	extinct	marine

Classification into higher taxa for these groups has been and remains problematic. A phylogenetic study suggests that the *Polyplacophora* form a clade with a monophyletic Aplacophora.[50] Additionally it suggests that a sister taxon relationship exists between the Bivalvia and the Gastropoda.

Evolution

Fossil record

There is good evidence for the appearance of gastropods, cephalopods and bivalves in the Cambrian period **unknown operator: u'cambrian'** to 488.3 [2] million years ago. However, the evolutionary history both of the emergence of molluscs from the ancestral group Lophotrochozoa, and of their diversification into the well-known living and fossil forms, is still vigorously debated.

There is debate about whether some Ediacaran and Early Cambrian fossils really are molluscs. *Kimberella*, from about 555 [51] million years ago, has been described as "mollusc-like",[52] [53] but others are unwilling to go further than "probable bilaterian".[54] [55] There is an even sharper debate about whether *Wiwaxia*, from about 505 [56] million years ago, was a mollusc, and much of this centers on whether its feeding apparatus was a type of radula or more similar to that of some polychaete worms.[54] [57] Nicholas Butterfield, who opposes the idea that *Wiwaxia* was a mollusc, has written that earlier microfossils from 515 to 510.0 [58] million years ago are fragments of a genuinely mollusc-like radula.[59] This appears to contradict the concept that the ancestral molluscan radula was mineralized.[60]

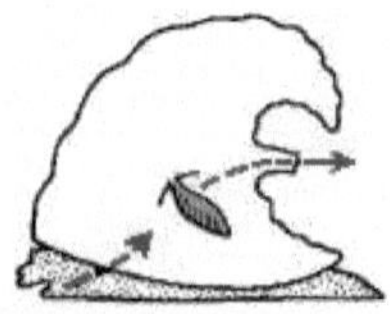

The tiny Helcionellid fossil *Yochelcionella* is thought to be an early mollusc[49]

Spirally coiled shells appear in many gastropods[13]

However, the Helcionellids, which first appear over 540 [61] million years ago in Early Cambrian rocks from Siberia and China,[62] [63] are thought to be early molluscs with rather snail-like shells. Shelled molluscs therefore predate the earliest trilobites.[49] Although most helcionellid fossils are only a few millimeters long, specimens a few centimeters long have also been found, most with more limpet-like shapes. There have been suggestions that the tiny specimens were juveniles and the larger ones adults.[64]

Some analyses of helcionellids concluded that these were the earliest gastropods.[65] However other scientists are not convinced that Early Cambrian fossils show clear signs of the torsion that identifies modern gastropods twists the internal organs so that the anus lies above the head.[13] [66] [67]

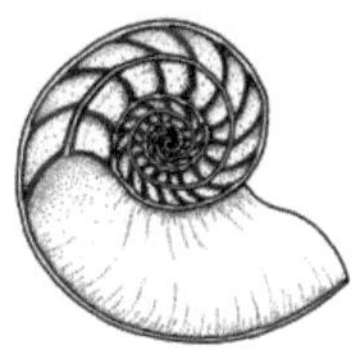

= Septa

= Siphuncle

Septa and siphuncle in nautiloid shell

For a long time it was thought that *Volborthella*, some fossils of which pre-date 530 [68] million years ago, was a cephalopod. However discoveries of more detailed fossils showed that *Volborthella*'s shell was not secreted but built from grains of the mineral silicon dioxide (silica), and that it was not divided into a series of compartments by septa as those of fossil shelled cephalopods and the living *Nautilus* are. *Volborthella*'s classification is uncertain.[69] The Late Cambrian fossil *Plectronoceras* is now thought to be the earliest clearly cephalopod fossil, as its shell had septa and a siphuncle, a strand of tissue that *Nautilus* uses to remove water from compartments that it has vacated as it grows, and which is also visible in fossil ammonite shells. However, *Plectronoceras* and other early cephalopods crept along the seafloor instead of swimming, as their shells contained a "ballast" of stony deposits on what is thought to be the underside and had stripes and blotches on what is thought to be the upper surface.[70] All cephalopods with external shells except the nautiloids became extinct by the end of the Cretaceous period 65 [71] million years ago.[72] However, the shell-less Coleoidea (squid, octopus, cuttlefish) are abundant today.[73]

The Early Cambrian fossils *Fordilla* and *Pojetaia* are regarded as bivalves.[74] [75] [76] [77] "Modern-looking" bivalves appeared in the Ordovician period, 488 to 443 [78] million years ago.[79] One bivalve group, the rudists, became major reef-builders in the Cretaceous, but became extinct in the Cretaceous-Tertiary extinction.[80] Even so, bivalves remain abundant and diverse.

The Hyolitha is a class of extinct animals with a shell and operculum that may be molluscs. Authors who suggest that they deserve their own phylum do not comment on the position of this phylum in the tree of life[81]

Phylogeny

Brachiopods

Bivalves

Monoplacophorans
("limpet-like", "living fossils")

Gastropods
(snails, slugs, limpets, sea hares)

Cephalopods
(nautiloids, ammonites, squid, etc.)

Scaphopods (tusk shells)

Lophotrochozoa

Aplacophorans
(spicule-covered, worm-like)

Polyplacophorans (chitons)

Halwaxiids   *Wiwaxia*

Halkieria

Orthrozanclus

Odontogriphus

A possible "family tree" of molluscs (2007).[82] [83] Does not include annelid worms as the analysis concentrated on fossilizable "hard" features.[82]

The phylogeny (evolutionary "family tree") of molluscs is a controversial subject. In addition to the debates about whether *Kimberella* and any of the "halwaxiids" were molluscs or closely related to molluscs,[53] [54] [57] [59] there are debates about the relationships between the classes of living molluscs.[55] In fact some groups traditionally classifed as molluscs may have to be redefined as distinct but related.[84]

Molluscs are generally regarded members of the Lophotrochozoa,[82] a group defined by having trochophore larvae and, in the case of living Lophophorata, a feeding structure called a lophophore. The other members of the Lophotrochozoa are the annelid worms and seven marine phyla.[85] The diagram on the right summarizes a phylogeny presented in 2007.

Because the relationships between the members of the family tree are uncertain, it is difficult to identify the features inherited from the last common ancestor of all molluscs.[86] For example, it is uncertain whether the ancestral mollusc was metameric (composed of repeating units)—if it was, that would suggest an origin from an annelid-like worm.[87] Scientists disagree about this: Giribet and colleagues concluded in 2006 that the repetition of gills and of the foot's retractor muscles were later developments, [6] while in 2007 Sigwart concluded that the ancestral mollusc was metameric, and that it had a foot used for creeping and a "shell" that was mineralized.[55] In one particular one branch of the family tree, the shell of conchiferans is thought to have evolved from the spicules (small spines) of aplacophorans; however this is difficult to reconcile with the embryological origins of spicules.[86]

The molluscan shell appears to have originated from a mucus coating, which eventually stiffened into a cuticle. This would have been impermeable and thus forced the development of more sophisticated respiratory apparatus in the form of gills.[49] Eventually, the cuticle would have become mineralized,[49] using the same genetic machinery (*engrailed*) as most other bilaterian skeletons.[87] The first mollusc shell almost certainly was reinforced with the mineral aragonite.[88]

The evolutionary relationships *within* the molluscs are also debated, and the diagrams below show two widely supported reconstructions:

<table>
<tr><td colspan="3"></td><td>Solenogastres</td></tr>
<tr><td></td><td>Solenogastres</td><td></td><td>Caudofoveata</td></tr>
<tr><td>Aculifera</td><td>Caudofoveata</td><td></td><td></td></tr>
<tr><td></td><td></td><td></td><td>Polyplacophorans</td></tr>
<tr><td></td><td>Polyplacophorans</td><td></td><td></td></tr>
<tr><td></td><td></td><td></td><td>Monoplacophorans</td></tr>
<tr><td>Molluscs</td><td>Monoplacophorans</td><td>Molluscs</td><td></td></tr>
<tr><td></td><td></td><td>Testaria</td><td>Bivalves</td></tr>
<tr><td></td><td>Bivalves</td><td></td><td>Scaphopods</td></tr>
<tr><td>Conchifera</td><td>Scaphopods</td><td></td><td>Gastropods</td></tr>
<tr><td></td><td>Gastropods</td><td></td><td>Cephalopods</td></tr>
<tr><td></td><td>Cephalopods</td><td></td><td></td></tr>
</table>

The "Aculifera" hypothesis[82]   The "Testaria" hypothesis[82]

Morphological analyses tend to recover a conchiferan clade that receives less support from molecular analyses,[89] although these results also lead to unexpected paraphylies, for instance scattering the bivalves throughout all other mollusc groups.[90]

However, an analysis in 2009 that used both morphological and molecular phylogenetics comparisons concluded that the molluscs are not monophyletic; in particular, that Scaphopoda and Bivalvia are both separate, monophyletic lineages unrelated to the remaining molluscan classes—in other words that the traditional phylum Mollusca is polyphyletic, and that it can only be made monophyletic if scaphopods and bivalves are excluded.[84] A 2010 analysis managed to recover the traditional conchiferan and aculiferan groups, but similarly concluded that the molluscs are not monophyletic, this time suggesting that solenogastres are more closely related to the non-molluscan taxa used as an outgroup than to other molluscs.[91] Current molecular data is insufficient to constrain the molluscan phylogeny, and since the methods used to determine the confidence in clades are prone to over-estimation, it is risky to place too much emphasis even on the areas that different studies agree.[92] Rather than eliminating unlikely relationships, the latest studies add new permutations of internal molluscan relationships, even bringing the conchiferan hypothesis into question.[93] [94]

Human interaction

For millennia molluscs have been a source of food for humans, as well as important luxury goods, notably pearls, mother of pearl, Tyrian purple dye, sea silk, and chemical compounds. Their shells have also been used as a form of currency in some pre-industrial societies. Their outlandish forms have helped conjure up tales of mythological sea monsters such as the Kraken. A number of species of molluscs can bite or sting humans, and some have become agricultural pests.

Uses by humans

Further information: Seashell

Molluscs, especially bivalves such as clams and mussels, have been an important food source since at least the advent of anatomically modern humans—and this has often resulted in over-fishing.[95] Other commonly eaten molluscs include octopuses and squids, whelks, oysters, and scallops.[96] In 2005, China accounted for 80% of the global mollusc catch, netting almost 11000000 tonnes (long tons; short tons). Within Europe, France remained the industry leader.[97] Some countries regulate importation and handling of molluscs and other seafood, mainly to minimize the poison risk from toxins that accumulate in the animals.[98]

Saltwater pearl oyster farm in Seram, Indonesia

Most molluscs that have shells can produce pearls, but only the pearls of bivalves and some gastropods whose shells are lined with nacre are valuable.[13] [15] The best natural pearls are produced by marine pearl oysters. *Pinctada margaritifera* and *Pinctada mertensi*, which live in the tropical and sub-tropical waters of the Pacific Ocean. Natural pearls form when a small foreign object gets stuck between the mantle and shell.

There are two methods of culturing pearls, by inserting either "seeds" or beads into oysters. The "seed" method uses grains of ground shell from freshwater mussels, and over-harvesting for this purpose has endangered several freshwater mussel species in the southeastern USA.[15] The pearl industry is so important in some areas that significant sums of money are spent on monitoring the health of farmed molluscs.[99]

Other luxury and high-status products were made from molluscs. Tyrian purple, made from the ink glands of murex shells, "... fetched its weight in silver" in the fourth-century BC, according to Theopompus.[100] The discovery of large numbers of Murex shells on Crete suggests that the Minoans may have pioneered the extraction of "Imperial purple" during the Middle Minoan period in the 20th–18th century BC, centuries before the Tyrians.[101] [102] Sea silk is a fine, rare and valuable fabric produced from the long silky threads (byssus) secreted by several bivalve molluscs, particularly *Pinna nobilis*, to attach themselves to the sea bed.[103] Procopius, writing on the Persian wars circa 550 CE, "stated that the five hereditary satraps (governors) of Armenia who received their insignia from the Roman Emperor were given chlamys (or cloaks) made from *lana pinna* (Pinna "wool," or byssus). Apparently only the ruling classes were allowed to wear these chlamys."[104]

Byzantine Emperor Justinian I clad in Tyrian purple and wearing numerous pearls

Mollusc shells, including those of cowries, were used as a kind of money (shell money) in several pre-industrial societies. However these "currencies" generally differed in important ways from the standardized government-backed and -controlled money familiar to industrial societies. Some shell "currencies" were not used for commercial transactions but mainly as social status displays at important occasions such as weddings.[105] When used for commercial transactions they functioned as commodity money, in other words as a tradable commodity whose value differed from place to place, often as a result of difficulties in transport, and which was vulnerable to incurable inflation if more efficient transport or "goldrush" behavior appeared.[106]

Stings and bites

There is a risk of food poisoning from toxins that accumulate in molluscs under certain conditions, and many countries have regulations that aim to minimize this risk. Blue-ringed octopus bites are often fatal, and the bite of other octopuses can cause unpleasant symptoms. Stings from a few species of large tropical cone shells can also kill. However, the sophisticated venoms of these cone snails have become important tools in neurological research and show promise as sources of new medications.

When handled alive, a few species of molluscs can sting or bite and, with some species, this can present a serious risk to the human handling the animal. To put this into perspective however, deaths from mollusc venoms are less than 10% of the number of deaths from jellyfish stings.[108]

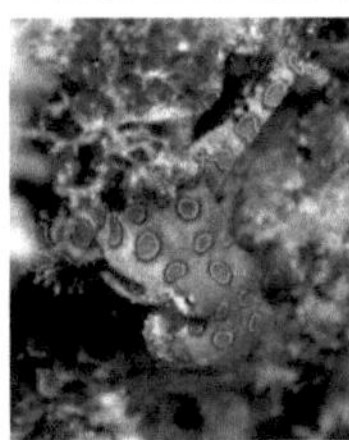

The blue-ringed octopus's rings are a warning signal—this octopus is alarmed, and its bite can kill.[107]

All octopuses are venomous[109] but only a few species pose a significant threat to humans. Blue-ringed octopuses in the genus *Hapalochlaena*, which live around Australia and New Guinea, bite humans only if severely provoked,[107] but their venom kills 25% of human victims. Another tropical species, *Octopus apollyon*, causes severe inflammation that can last for over a month even if treated correctly,[110] and the bite of *Octopus rubescens* can cause necrosis that lasts longer than one month if untreated, and headaches and weakness persisting for up to a week even if treated.[111]

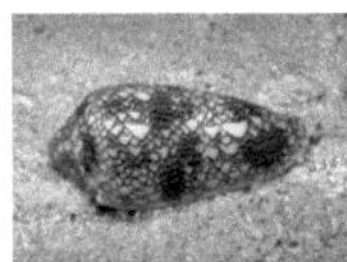

Live cone snails can be dangerous to shell-collectors but are useful to neurology researchers[112]

All species of cone snails are venomous and can sting when handled, although many species are too small to pose much of a risk to humans. These are carnivorous gastropods that feed on marine invertebrates (and in the case of larger species on fish). Their venom is based on a huge array of toxins, some fast-acting and others slower but deadlier—they can afford to do this because their toxins require less time and energy to be produced compared with those of snakes or spiders.[112] Many painful stings have been reported, and a few fatalities, although some of the reported fatalities may be exaggerations.[108] Only the few larger species of cone snail that can capture and kill fish are likely to be seriously dangerous to humans.[113] The effects of individual cone shell toxins on victims' nervous systems are so precise that they are useful tools for research in neurology, and the small size of their molecules makes it easy to synthesize them.[112] [114]

The traditional belief that a giant clam can trap the leg of a person between its valves, thus drowning them, is a myth.[115]

Pests

Schistosomiasis (also known as bilharzia, bilharziosis or snail fever) is transmitted to humans via water snail hosts, and affects about 200 million people. A few species of snails and slugs are serious agricultural pests, and in addition, accidental or deliberate introduction of various snail species into new territory has resulted in serious damage to some natural ecosystems.

Schistosomiasis is "second only to malaria as the most devastating parasitic disease in tropical countries. An estimated 200 million people in 74 countries are infected with the disease — 100 million in Africa alone."[116] The parasite has 13 known species, of which two infect humans. The parasite itself is not a mollusc, but all the species have freshwater snails as intermediate hosts.[117]

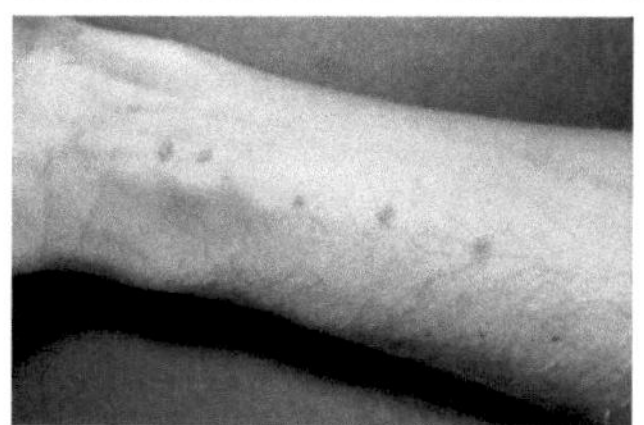

Skin vesicles created by the penetration of *Schistosoma*. Source: Centers for Disease Control and Prevention

Some species of molluscs, particularly certain snails and slugs, can be serious crop pests,[118] and when introduced into new environments can unbalance local ecosystems. One such pest, the giant African snail *Achatina fulica*, has been introduced to many parts of Asia, as well as to many islands in the Indian Ocean and Pacific Ocean. In the 1990s this species reached the West Indies. Attempts to control it by introducing the predatory snail *Euglandina rosea* proved disastrous, as the predator ignored *Achatina fulica* and went on to extirpate several native snail species instead.[119]

Despite its name, *Molluscum contagiosum* is a viral disease, and is unrelated to molluscs.[120]

Notes

Footnotes

[1] Spelled *mollusks* in the USA, see reasons given in Rosenberg's (http://www.conchologistsofamerica.org/articles/y1996/9609_rosenberg. asp); for the spelling *mollusc* see the reasons given by Brusca & Brusca. *Invertebrates* (2nd ed.).

[2] http://toolserver.org/~verisimilus/Timeline/Timeline.php?Ma=Cambrian–488.3

[3] Little, L., Fowler, H.W., Coulson, J., and Onions, C.T., ed (1964). "Mollusca". *Shorter Oxford English Dictionary*. Oxford University press.

[4] Little, L., Fowler, H.W., Coulson, J., and Onions, C.T., ed (1964). "Malacology". *Shorter Oxford English Dictionary*. Oxford University press.

[5] C. Michael Hogan. 2010. *Calcium*. eds. A.Jorgensen, C. Cleveland. Encyclopedia of Earth (http://www.eoearth.org/article/ Calcium?topic=49557). National Council for Science and the Environment.

[6] Giribet, G., Okusu, A., Lindgren, A.R., Huff, S.W., Schrödl, M., and Nishiguchi, M.K. (May 2006). "Evidence for a clade composed of molluscs with serially repeated structures: Monoplacophorans are related to chitons" (http://www.pnas.org/content/103/20/7723.full). *Proceedings of the National Academy of Sciences of the United States of America* **103** (20): 7723–7728. Bibcode 2006PNAS..103.7723G. doi:10.1073/pnas.0602578103. PMC 1472512. PMID 16675549. . Retrieved 2008-09-30.

[7] Hayward, PJ (1996). *Handbook of the Marine Fauna of North-West Europe*. Oxford University Press. pp. 484–628. ISBN 0198540558.

[8] Brusca, R.C., and Brusca, G.J. (2003). *Invertebrates* (2 ed.). Sinauer Associates. pp. 702. ISBN 0878930973.

[9] Ruppert, E.E., Fox, R.S., and Barnes, R.D. (2004). *Invertebrate Zoology* (7 ed.). Brooks / Cole. pp. 284–291. ISBN 0030259827.

[10] Ruppert, E.E., Fox, R.S., and Barnes, R.D. (2004). *Invertebrate Zoology* (7 ed.). Brooks / Cole. pp. 291–292. ISBN 0030259827.

[11] Ruppert, E.E., Fox, R.S., and Barnes, R.D. (2004). *Invertebrate Zoology* (7 ed.). Brooks / Cole. pp. 292–298. ISBN 0030259827.

[12] Ruppert, E.E., Fox, R.S., and Barnes, R.D. (2004). *Invertebrate Zoology* (7 ed.). Brooks / Cole. pp. 298–300. ISBN 0030259827.

[13] Ruppert, E.E., Fox, R.S., and Barnes, R.D. (2004). *Invertebrate Zoology* (7 ed.). Brooks / Cole. pp. 300–343. ISBN 0030259827.

[14] Ruppert, E.E., Fox, R.S., and Barnes, R.D. (2004). *Invertebrate Zoology* (7 ed.). Brooks / Cole. pp. 343–367. ISBN 0030259827.

[15] Ruppert, E.E., Fox, R.S., and Barnes, R.D. (2004). *Invertebrate Zoology* (7 ed.). Brooks / Cole. pp. 367–403. ISBN 0030259827.

[16] Ruppert, E.E., Fox, R.S., and Barnes, R.D. (2004). *Invertebrate Zoology* (7 ed.). Brooks / Cole. pp. 403–407. ISBN 0030259827.

[17] edited by Winston F. Ponder, David R. Lindberg. (2008). Ponder, W.F. and Lindberg, D.R.. ed. *Phylogeny and Evolution of the Mollusca*. Berkeley: University of California Press. pp. 481. ISBN 978-0520250925.

[18] Chapman, A.D. (2009). Numbers of Living Species in Australia and the World, 2nd edition (http://www.environment.gov.au/ biodiversity/abrs/publications/other/species-numbers/2009/04-02-groups-invertebrates.html#mollusca). Australian Biological Resources Study, Canberra. Accessed 12 January 2010. ISBN 978 0 642 56860 1 (printed); ISBN 978 0 642 56861 8 (online).

[19] Haszprunar, G. (2001). "Mollusca (Molluscs)". *Encyclopedia of Life Sciences*. John Wiley & Sons, Ltd.. doi:10.1038/npg.els.0001598.

[20] Rebecca Hancock (2008). "Recognising research on molluscs" (http://www.austmus.gov.au/display.cfm?id=2897). Australian Museum. . Retrieved 2009-03-09.

[21] Ruppert, E.E., Fox, R.S., and Barnes, R.D. (2004). *Invertebrate Zoology* (7 ed.). Brooks / Cole. pp. Front endpaper 1. ISBN 0030259827.

[22] Winston F. Ponder and David R. Lindberg (2004). "Phylogeny of the Molluscs" (http://www.ucmp.berkeley.edu/museum/news/news_briefs/perth08042004.html). World Congress of Malacology. . Retrieved 2009-03-09.

[23] David M. Raup & Steven M. Stanley (1978). *Principles of Paleontology* (2 ed.). W.H. Freeman and Co.. pp. 4–5. ISBN 071670220.

[24] Barnes, R.S.K., Calow, P., Olive, P.J.W., Golding, D.W. and Spicer, J.I. (2001). *The Invertebrates, A Synthesis* (3 ed.). UK: Blackwell Science.

[25] Kubodera, T. and Mori, K. (2005). "First-ever observations of a live giant squid in the wild" (http://www.canarias7.es/pdf/docs/informecalamargigante.pdf). *Proceedings of the Royal Society B: Biological Sciences* **272** (1581): 2583–2586. doi:10.1098/rspb.2005.3158. PMC 1559985. PMID 16321779. . Retrieved 2008-10-22.

[26] Richard Black (April 26, 2008). "Colossal squid out of the freezer" (http://news.bbc.co.uk/1/hi/sci/tech/7367774.stm). BBC News. . Retrieved 2008-10-01.

[27] Lydeard, C.; R. Cowie, R., Ponder, W.F., *et al* (April 2004). "The global decline of nonmarine mollusks" (http://www.unc.edu/~keperez/lydeard_bioscience.pdf). *BioScience* **54**: 321–330. doi:10.1641/0006-3568(2004)054[0321:TGDONM]2.0.CO;2. . Retrieved 20 Oct 2009.

[28] Ruppert, E.E., Fox, R.S., and Barnes, R.D. (2004). *Invertebrate Zoology* (7 ed.). Brooks / Cole. pp. 284–291. ISBN 0030259827.

[29] Healy, J.M. (2001). "The Mollusca". In Anderson, D.T.. *Invertebrate Zoology* (2 ed.). Oxford University Press. pp. 120–171. ISBN 0195513681.

[30] Porter, S. (2007). "Seawater Chemistry and Early Carbonate Biomineralization" (http://www.sciencemag.org/cgi/content/abstract/sci;316/5829/1302). *Science* **316** (5829): 1302. Bibcode 2007Sci...316.1302P. doi:10.1126/science.1137284. PMID 17540895. . Retrieved 2008-09-30.

[31] Yochelson, E. L. (1975). "Discussion of early Cambrian "molluscs"" (http://jgs.geoscienceworld.org/cgi/reprint/131/6/661.pdf). *Journal of the Geological Society* **131** (6): 661–662. doi:10.1144/gsjgs.131.6.0661. .

[32] Cherns, L. (2004). "Early Palaeozoic diversification of chitons (Polyplacophora, Mollusca) based on new data from the Silurian of Gotland, Sweden". *Lethaia* **37** (4): 445–456. doi:10.1080/00241160410002180.

[33] Tompa, A. S. (1976). "A comparative study of the ultrastructure and mineralogy of calcified land snail eggs (Pulmonata: Stylommatophora)". *Journal of Morphology* **150** (4): 861–887. doi:10.1002/jmor.1051500406.

[34] Wilbur, Karl M.; Trueman, E.R.; Clarke, M.R., eds. (1985), *The Mollusca*, **11. Form and Function**, New York: Academic Press, ISBN 0-12-728702-7

[35] Shigeno, S; Sasaki, T; Moritaki, T; Kasugai, T; Vecchione, M; Agata, K (Jan 2008). "Evolution of the cephalopod head complex by assembly of multiple molluscan body parts: Evidence from Nautilus embryonic development.". *Journal of morphology* **269** (1): 1–17. doi:10.1002/jmor.10564. PMID 17654542.

[36] Ruppert, E.E., Fox, R.S., and Barnes, R.D. (2004). "Mollusca". *Invertebrate Zoology* (7 ed.). Brooks / Cole. pp. 290–291. ISBN 0030259827.

[37] Marin, F.; Luquet, G. (2004). "Molluscan shell proteins". *Comptes Rendus Palevol* **3** (6-7): 469. doi:10.1016/j.crpv.2004.07.009.

[38] Steneck, R. S.; Watling, L. (1982). "Feeding capabilities and limitation of herbivorous molluscs: A functional group approach". *Marine Biology* **68** (3): 299–319. doi:10.1007/BF00409596.

[39] *Galathea* **16**: 095–098. http://www.zmuc.dk/InverWeb/Galathea/Pdf_filer/Volume_16/galathea-vol.16-pp_095-098.pdf.

[40] Jensen, K. R. (1993). "Morphological adaptations and plasticity of radular teeth of the Sacoglossa (= Ascoglossa) (Mollusca: Opisthobranchia) in relation to their food plants". *Biological Journal of the Linnean Society* **48** (2): 135–155. doi:10.1111/j.1095-8312.1993.tb00883.x.

[41] W◆Gele, H. (1989). "Diet of some Antarctic nudibranchs (Gastropoda, Opisthobranchia, Nudibranchia)". *Marine Biology* **100** (4): 439–441. doi:10.1007/BF00394819.

[42] Publishers, Bentham Science (1999-07). *Current Organic Chemistry* (http://books.google.com/?id=0HO-3M534nEC&pg=PA327). .

[43] . http://www.biolbull.org/cgi/content/abstract/181/2/248.

[44] http://mollus.oxfordjournals.org/content/31/3-4/144.short

[45] Ruppert, E.E., Fox, R.S., and Barnes, R.D. (2004). *Invertebrate Zoology* (7 ed.). Brooks / Cole. pp. 300. ISBN 0030259827.

[46] Ruppert, E.E., Fox, R.S., and Barnes, R.D. (2004). *Invertebrate Zoology* (7 ed.). Brooks / Cole. pp. 343. ISBN 0030259827.

[47] Ruppert, E.E., Fox, R.S., and Barnes, R.D. (2004). *Invertebrate Zoology* (7 ed.). Brooks / Cole. pp. 367. ISBN 0030259827.

[48] Clarkson, E.N.K., (1998). *Invertebrate Palaeontology and Evolution* (http://books.google.com/?id=g1P2VaPQWfUC&pg=PA221&dq="Invertebrate+Palaeontology+and+Evolution"+rostroconchia#PPA221,M1). Blackwell. pp. 221. ISBN 0632052384. . Retrieved 2008-10-27.

[49] Runnegar, B. and Pojeta, J. (1974). "Molluscan phylogeny: the paleontological viewpoint" (http://www.sciencemag.org/cgi/content/abstract/186/4161/311). *Science* **186** (4161): 311–7. Bibcode 1974Sci...186..311R. doi:10.1126/science.186.4161.311. PMID 17839855. . Retrieved 2008-07-30.

[50] Kocot KM, Cannon JT, Todt C, Citarella MR, Kohn AB, Meyer A, Santos SR, Schander C, Moroz LL, Lieb B, Halanych KM (2011) Phylogenomics reveals deep molluscan relationships. Nature doi: 10.1038/nature10382.

[51] http://toolserver.org/~verisimilus/Timeline/Timeline.php?Ma=555

[52] Fedonkin, M.A.; Waggoner, B.M. (1997). "The Late Precambrian fossil Kimberella is a mollusc-like bilaterian organism". *Nature* **388** (6645): 868. Bibcode 1997Natur.388..868F. doi:10.1038/42242.

[53] Fedonkin, M.A., Simonetta, A. and Ivantsov, A.Y. (2007). "New data on Kimberella, the Vendian mollusc-like organism (White Sea region, Russia): palaeoecological and evolutionary implications" (http://www.geosci.monash.edu.au/precsite/docs/workshop/prato04/abstracts/ fedonkin2.pdf). *Geological Society, London, Special Publications* **286**: 157–179. doi:10.1144/SP286.12. . Retrieved 2008-07-10.

[54] Butterfield, N.J. (2006). "Hooking some stem-group "worms": fossil lophotrochozoans in the Burgess Shale". *Bioessays* **28** (12): 1161–6. doi:10.1002/bies.20507. PMID 17120226.

[55] Sigwart; Sutton, M. D. (Oct 2007). "Deep molluscan phylogeny: synthesis of palaeontological and neontological data". *Proceedings of the Royal Society B: Biological sciences* **274** (1624): 2413–2419. doi:10.1098/rspb.2007.0701. PMC 2274978. PMID 17652065. For a summary, see "The Mollusca" (http://www.ucmp.berkeley.edu/taxa/inverts/mollusca/mollusca.php). University of California Museum of Paleontology. . Retrieved 2008-10-02.

[56] http://toolserver.org/~verisimilus/Timeline/Timeline.php?Ma=505

[57] Caron, J.B.; Scheltema, A., Schander, C., and Rudkin, D. (2006-07-13). "A soft-bodied mollusc with radula from the Middle Cambrian Burgess Shale" (http://www.nature.com/nature/journal/v442/n7099/pdf/nature04894.pdf). *Nature* **442** (7099): 159–163. Bibcode 2006Natur.442..159C. doi:10.1038/nature04894. PMID 16838013. . Retrieved 2008-08-07.

[58] http://toolserver.org/~verisimilus/Timeline/Timeline.php?Ma=515–510

[59] Butterfield, N.J. (May 2008). "An Early Cambrian Radula" (http://findarticles.com/p/articles/mi_qa3790/is_200805/ai_n25501673/ pg_1?tag=artBody;coll). *Journal of Paleontology* **82** (3): 543–554. doi:10.1666/07-066.1. . Retrieved 2008-08-20.

[60] Cruz, R.; Lins, U.; Farina, M. (1998). "Minerals of the radular apparatus of *Falcidens* sp. (Caudofoveata) and the evolutionary implications for the Phylum Mollusca". *Biological Bulletin* **194** (2): 224–230. doi:10.2307/1543051. JSTOR 1543051.

[61] http://toolserver.org/~verisimilus/Timeline/Timeline.php?Ma=540

[62] P. Yu. Parkhaev (2007). "The Cambrian 'basement' of gastropod evolution" (http://books.google.com/?id=GA7-8JIh9IwC&pg=PA415). *Geological Society, London, Special Publications* **286**: 415–421. doi:10.1144/SP286.31. ISBN 9781862392335. . Retrieved 2009-11-01.

[63] Michael Steiner, Guoxiang Li, Yi Qian, Maoyan Zhu and Bernd-Dietrich Erdtmann (2007). "Neoproterozoic to early Cambrian small shelly fossil assemblages and a revised biostratigraphic correlation of the Yangtze Platform (China)". *Palaeogeography, Palaeoclimatology, Palaeoecology* **254**: 67–99. doi:10.1016/j.palaeo.2007.03.046.

[64] Mus, M. M.; Palacios, T.; Jensen, S. (2008). "Size of the earliest mollusks: Did small helcionellids grow to become large adults?" (http:// geology.geoscienceworld.org/cgi/content/abstract/36/2/175). *Geology* **36** (2): 175. doi:10.1130/G24218A.1. . Retrieved 2008-10-01.

[65] Landing, E. .; Geyer, G. .; Bartowski, K. E. (March 2002). "Latest Early Cambrian Small Shelly Fossils, Trilobites, and Hatch Hill Dysaerobic Interval on the Quebec Continental Slope". *Journal of Paleontology* **76** (2): 287–305. doi:10.1666/0022-3360(2002)076<0287:LECSSF>2.0.CO;2. ISSN 0022-3360.

[66] Frýda, J.; Nützel, A. and Wagner, P.J. (2008). "Paleozoic Gastropoda" (http://books.google.com/?id=nm0IZAQQ6S0C& pg=RA1-PA239&dq=earliest+gastropod+fossil#v=onepage&q=earliest gastropod fossil). In Ponder, W.F., and Lindberg, D.R.. *Phylogeny and evolution of the Mollusca*. California Press. pp. 239–264. ISBN 0520250923. . Retrieved 4 Nov 2009.

[67] Kouchinsky, A. (2000). "Shell microstructures in Early Cambrian molluscs" (http://app.pan.pl/archive/published/app45/app45-119. pdf). *Acta Palaeontologica Polonica* **45** (2): 119–150. . Retrieved 4 Nov 2009.

[68] http://toolserver.org/~verisimilus/Timeline/Timeline.php?Ma=530

[69] Hagadorn, J.W., and Waggoner, B.M. (2002). "The Early Cambrian problematic fossil Volborthella: New insights from the Basin and Range" (http://www3.amherst.edu/~jwhagadorn/publications/volb.pdf). In Corsetti, F.A.. *Proterozoic-Cambrian of the Great Basin and Beyond, Pacific Section SEPM Book 93*. SEPM (Society for Sedimentary Geology). pp. 135–150. . Retrieved 2008-10-01.

[70] Vickers-Rich, P., Fenton, C.L., Fenton, M.A. and Rich, T.H. (1997). *The Fossil Book: A Record of Prehistoric Life*. Courier Dover Publications. pp. 269–272. ISBN 0486293718.

[71] http://toolserver.org/~verisimilus/Timeline/Timeline.php?Ma=65

[72] Marshall C.R., and Ward P.D. (1996). "Sudden and Gradual Molluscan Extinctions in the Latest Cretaceous of Western European Tethys". *Science* **274** (5291): 1360–1363. Bibcode 1996Sci...274.1360M. doi:10.1126/science.274.5291.1360. PMID 8910273.

[73] Monks, N.. "A Broad Brush History of the Cephalopoda" (http://www.thecephalopodpage.org/evolution.php). . Retrieved 2009-03-21.

[74] Pojeta, J. (2000). "Cambran Pelecypoda (Mollusca)". *American Malacological Bulletin* **15**: 157–166.

[75] Schneider, J.A. (November 2001). "Bivalve systematics during the 20th century" (http://findarticles.com/p/articles/mi_qa3790/ is_200111/ai_n9001371/pg_3). *Journal of Paleontology* **75** (6): 1119–1127. doi:10.1666/0022-3360(2001)075<1119:BSDTC>2.0.CO;2. . Retrieved 2008-10-05.

[76] Gubanov, A.P., Kouchinsky, A.V. and Peel, J.S. (2007). "The first evolutionary-adaptive lineage within fossil molluscs". *Lethaia* **32** (2): 155–157. doi:10.1111/j.1502-3931.1999.tb00534.x.

[77] Gubanov, A.P., and Peel, J.S. (2003). "The early Cambrian helcionelloid mollusc *Anabarella* Vostokova". *Palaeontology* **46** (5): 1073–1087. doi:10.1111/1475-4983.00334.

[78] http://toolserver.org/~verisimilus/Timeline/Timeline.php?Ma=488–443

[79] Zong-Jie, F. (2006). "An introduction to Ordovician bivalves of southern China, with a discussion of the early evolution of the Bivalvia". *Geological Journal* **41** (3-4): 303–328. doi:10.1002/gj.1048.

[80] Raup, D.M., and Jablonski, D. (1993). "Geography of end-Cretaceous marine bivalve extinctions". *Science* **260** (5110): 971–973. Bibcode 1993Sci...260..971R. doi:10.1126/science.11537491. PMID 11537491.

[81] Malinky, j. 2009 "Permian Hyolithida from Australia: The Last of the Hyoliths?" *Journal of Paleontology* 83(1):147–152.

[82] Sigwart, J.D., and Sutton, M.D. (October 2007). "Deep molluscan phylogeny: synthesis of palaeontological and neontological data". *Proceedings of the Royal Society: Biology* **274** (1624): 2413–2419. doi:10.1098/rspb.2007.0701. PMC 2274978. PMID 17652065. For a summary, see "The Mollusca" (http://www.ucmp.berkeley.edu/taxa/inverts/mollusca/mollusca.php). University of California Museum of Paleontology. . Retrieved 2008-10-02.

[83] "The Mollusca" (http://www.ucmp.berkeley.edu/taxa/inverts/mollusca/mollusca.php). University of California Museum of Paleontology. . Retrieved 2008-10-02.

[84] Goloboff, P. A.; Catalano, S. A.; Marcos Mirande, J.; Szumik, C. A.; Salvador Arias, J.; Källersjö, M.; Farris, J. S. (2009). "Phylogenetic analysis of 73 060 taxa corroborates major eukaryotic groups". *Cladistics* **25** (3): 211. doi:10.1111/j.1096-0031.2009.00255.x.

[85] "Introduction to the Lophotrochozoa" (http://www.ucmp.berkeley.edu/phyla/lophotrochozoa.html). University of California Museum of Paleontology. . Retrieved 2008-10-02.

[86] Henry, J.; Okusu, A.; Martindale, M. (2004). "The cell lineage of the polyplacophoran, Chaetopleura apiculata: variation in the spiralian program and implications for molluscan evolution". *Developmental biology* **272** (1): 145–160. doi:10.1016/j.ydbio.2004.04.027. PMID 15242797.

[87] Jacobs, D. K.; Wray, C. G.; Wedeen, C. J.; Kostriken, R.; Desalle, R.; Staton, J. L.; Gates, R. D.; Lindberg, D. R. (2000). "Molluscan engrailed expression, serial organization, and shell evolution". *Evolution & Development* **2** (6): 340–347. doi:10.1046/j.1525-142x.2000.00077.x. PMID 11256378.

[88] Porter, S. M. (Jun 2007). "Seawater chemistry and early carbonate biomineralization". *Science* **316** (5829): 1302–1301. Bibcode 2007Sci...316.1302P. doi:10.1126/science.1137284. ISSN 0036-8075. PMID 17540895.

[89] Winnepenninckx, B; Backeljau, T; De Wachter, R (December 1, 1996). "Investigation of molluscan phylogeny on the basis of 18S rRNA sequences." (http://mbe.oxfordjournals.org/content/13/10/1306.abstract). *Molecular Biology and Evolution* **13** (10): 1306–1317. PMID 8952075. .

[90] Passamaneck, Y.; Schander, C.; Halanych, K. (2004). "Investigation of molluscan phylogeny using large-subunit and small-subunit nuclear rRNA sequences.". *Molecular phylogenetics and evolution* **32** (1): 25–38. doi:10.1016/j.ympev.2003.12.016. PMID 15186794.

[91] Wilson, N.; Rouse, G.; Giribet, G. (2010). "Assessing the molluscan hypothesis Serialia (Monoplacophora+Polyplacophora) using novel molecular data.". *Molecular phylogenetics and evolution* **54** (1): 187–193. doi:10.1016/j.ympev.2009.07.028. PMID 19647088.

[92] Wägele, J.; Letsch, H.; Klussmann-Kolb, A.; Mayer, C.; Misof, B.; Wägele, H. (2009). "Phylogenetic support values are not necessarily informative: the case of the Serialia hypothesis (a mollusk phylogeny)". *Frontiers in zoology* **6** (1): 12. doi:10.1186/1742-9994-6-12. PMC 2710323. PMID 19555513.

[93] Vinther, J.; Sperling, E. A.; Briggs, D. E. G.; Peterson, K. J. (2011). "A molecular palaeobiological hypothesis for the origin of aplacophoran molluscs and their derivation from chiton-like ancestors". *Proceedings of the Royal Society B: Biological Sciences*. doi:10.1098/rspb.2011.1773.

[94] doi:10.1038/nature1038
This citation will be automatically completed in the next few minutes. You can jump the queue or expand by hand (http://en.wikipedia.org/wiki/Template:cite_doi/10.1038.2fnature1038_)

[95] Mannino, M.A., and Thomas, K.D. (February 2002). "Depletion of a resource? The impact of prehistoric human foraging on intertidal mollusc communities and its significance for human settlement, mobility and dispersal". *World Archaeology* **33** (3): 452–474. doi:10.1080/00438240120107477.

[96] Garrow, J.S., Ralph, A., and James, W.P.T. (2000). *Human Nutrition and Dietetics*. Elsevier Health Sciences. pp. 370. ISBN 0443056277.

[97] "China catches almost 11 m tonnes of molluscs in 2005" (http://www.fao.org/figis/servlet/TabLandArea?tb_ds=Capture&tb_mode=TABLE&tb_act=SELECT&tb_grp=COUNTRY). FAO. . Retrieved 2008-10-03.

[98] "Importing fishery products or bivalve molluscs" (http://www.food.gov.uk/foodindustry/imports/want_to_import/fisheryproducts/). United Kingdom: Food Standards Agency. . Retrieved 2008-10-02.

[99] Jones, J.B., and Creeper, J. (April 2006). "Diseases of Pearl Oysters and Other Molluscs: a Western Australian Perspective". *Journal of Shellfish Research* **25** (1): 233–238. doi:10.2983/0730-8000(2006)25[233:DOPOAO]2.0.CO;2.

[100] The fourth-century BC historian Theopompus, cited by Athenaeus (12:526) around 200 BC ; according to Gulick, C.B. (1941). *Athenaeus, The Deipnosophists*. Cambridge, Mass.: Harvard University Press. ISBN 0674993802.

[101] Reese, D.S. (1987). "Palaikastro Shells and Bronze Age Purple-Dye Production in the Mediterranean Basin". *Annual of the British School of Archaeology at Athens* **82**: 201–6.

[102] Stieglitz, R.R. (1994). "The Minoan Origin of Tyrian Purple". *Biblical Archaeologist* **57** (1): 46–54. doi:10.2307/3210395. JSTOR 3210395.

[103] *Webster's Third New International Dictionary (Unabridged)* 1976. G. & C. Merriam Co., p. 307.

[104] Turner, R.D., and Rosewater, J. (June 1958). "The Family Pinnidae in the Western Atlantic". *Johnsonia* **3** (38): 294.

[105] Maurer, B. (October 2006). "The Anthropology of Money" (http://www.anthro.uci.edu/faculty_bios/maurer/Maurer-AR.pdf). *Annual Review of Anthropology* **35**: 15–36. doi:10.1146/annurev.anthro.35.081705.123127. . Retrieved 2008-10-23.

[106] Hogendorn, J., and Johnson, M. (2003). *The Shell Money of the Slave Trade*. Cambridge University Press. ISBN 052154110. Particularly chapters "Boom and slump for the cowrie trade" (pages 64–79) and "The cowrie as money: transport costs, values and inflation" (pages 125–147)

[107] Alafaci, A.. "Blue ringed octopus" (http://www.avru.org/compendium/biogs/A000060b.htm). Australian Venom Research Unit. . Retrieved 2008-10-03.

[108] Williamson, J.A., Fenner, P.J., Burnett, J.W., and Rifkin, J. (1996). *Venomous and Poisonous Marine Animals: A Medical and Biological Handbook* (http://books.google.com/?id=YsZ3GryFIzEC&pg=PA75&lpg=PA75&dq=mollusc+venom+fatal). UNSW Press. pp. 65–68. ISBN 0868402796. . Retrieved 2008-10-03.

[109] Anderson, R.C. (1995). "Aquarium husbandry of the giant Pacific octopus". *Drum and Croaker* **26**: 14–23.

[110] Brazzelli, V., Baldini, F., Nolli, G., Borghini, F., and Borroni, G. (1999). "*Octopus apollyon* bite". *Contact Dermatitis* **40** (3): 169–170. doi:10.1111/j.1600-0536.1999.tb06025.x. PMID 10073455.

[111] Anderson, R.C. (1999). "An octopus bite and its treatment". *The Festivus* **31**: 45–46.

[112] Concar, D. (19 October 1996). "Doctor snail—Lethal to fish and sometimes even humans, cone snail venom contains a pharmacopoeia of precision drugs" (http://environment.newscientist.com/article/mg15220523. 900-doctor-snail--lethal-to-fish-and-sometimes-even-humans-cone-snail-venom-contains-apharmacopoeia-of-precision-drugs-itdavid-concarit-finds-out-how-th html). *New Scientist*. . Retrieved 2008-10-03.

[113] Livett, B.. "Cone Shell Mollusc Poisoning, with Report of a Fatal Case" (http://grimwade.biochem.unimelb.edu.au/cone/deathby. html). Department of Biochemistry and Molecular Biology, University of Melbourne. .

[114] Haddad, V.(junior), de Paula Neto, J.B., and Cobo, V.J. (September–October 2006). "Venomous mollusks: the risks of human accidents by *Conus* snails (Gastropoda: Conidae) in Brazil" (http://www.scielo.br/pdf/rsbmt/v39n5/a15v39n5.pdf). *Revista da Sociedade Brasileira de Medicina Tropical* **39** ((5)): 498–500. . Retrieved 2008-10-03.

[115] Cerullo, M.M., Rotman, J.L., and Wertz, M. (2003). *The Truth about Dangerous Sea Creatures* (http://books.google.com/ ?id=1MOxNDmFLd4C&pg=PA1&dq=giant+clam+trap+foot). Chronicle Books. pp. 10. ISBN 0811840506. . Retrieved 2008-10-03.

[116] "The Carter Center Schistosomiasis Control Program" (http://www.cartercenter.org/health/schistosomiasis/index.html). The Carter Center. . Retrieved 2008-10-03.

[117] Brown, D.S. (1994). *Freshwater Snails of Africa and Their Medical Importance*. CRC Press. pp. 305. ISBN 0748400265.

[118] Barker, G.M. (2002). *Molluscs As Crop Pests*. CABI Publications. ISBN 0851993206.

[119] Civeyrel, L., and Simberloff, D. (October 1996). "A tale of two snails: is the cure worse than the disease?". *Biodiversity and Conservation* **5** (10): 1231–1252. doi:10.1007/BF00051574.

[120] "Molluscum (Molluscum Contagiosum): Frequently Asked Questions for Everyone" (http://www.cdc.gov/ncidod/dvrd/molluscum/ faq/everyone.htm). Centers for Disease Control and Prevention. . Retrieved 2008-10-03.

Citations

Further reading

- Starr & Taggart (2002). *Biology: The Unity and Diversity of Life*. Pacific Grove, California: Thomson Learning. ISBN 0534027423.
- Nunn, J.D., Smith, S.M., Picton, B.E. and McGrath, D. (2002). "Checklist, atlas of distribution and bibliography for the marine mollusca of Ireland". *Marine Biodiversity in Ireland and Adjacent Waters*. **8**. Ulster Museum.
- Ostroumov SA (2001). "An amphiphilic substance inhibits the mollusk capacity to filter phytoplankton cells from water.". *Izvestiia Akademii nauk. Seriia biologicheskaia / Rossiiskaia akademiia nauk* (1): 108–16. PMID 11236572.; http://www.springerlink.com/content/l665628020163255/;
- Dame, R.; Olenin, S., eds (2005). *The comparative roles of suspension-feeders in ecosystems*. Dordrecht: Springer.

External links

- Researchers complete mollusk evolutionary tree; 26 October 2011 (http://www.physorg.com/news/ 2011-10-mollusk-evolutionary-tree.html)
- Hardy's Internet Guide to Marine Gastropods (http://www.gastropods.com/)
- Planktonic mollusca fact sheets (http://www.tafi.org.au/zooplankton/imagekey/mollusca/index.html)
- Rotterdam Natural History Museum (http://www.nmr-pics.nl/) Shell Image Gallery

Sea_snail

Sea snail is a common name for those snails that normally live in saltwater, marine gastropod molluscs. (The taxonomic class Gastropoda also includes snails that live in other habitats, i.e. land snails and freshwater snails.)

Sea snails are marine gastropods that have shells. Those marine gastropods that have no shells, or have only internal shells, are variously known by other common names, including sea slug, sea hare, nudibranch, etc.

Many sea snails are edible and are exploited as food sources by humans. Some well-known kinds of edible sea snails are abalone, conch, limpets, whelks (such as the North American *Busycon* species and the North Atlantic *Buccinum undatum*) and periwinkles including *Littorina littorea*.

A species of sea snail in its natural habitat: two individuals of the wentletrap *Epidendrium billeeanum* with a mass of egg capsules *in situ* on their food source, a red cup coral

There is enormous diversity within sea snails; many very different clades of gastropods are either dominated by, or consist exclusively of, sea snails. Because of this great variability, it is not possible to generalize about the feeding, reproduction, habitat and so on of sea snails. Instead it is necessary to look at the articles about individual clades, families, genera or species.

Shells

The shells of most species of sea snails are spirally coiled; some however have shells that are conical, and these are often referred to by the common name of limpets. In one unusual family Juliidae, the shell of the snail has become two hinged plates closely resembling those of a bivalve; this family is sometimes called the "bivalved gastropods".

The shells of living species of sea snails range in size from *Syrinx aruanus*, the largest living shelled gastropod species, to minute species whose shells are under 1 mm at adult size.

Because in many cases the shells of sea snails are strong and durable, as a group they are well represented in the fossil record.

A group of *Patella vulgata* limpets on a rock in Pembrokeshire

Anatomy

Sea snails are a very large group of animals and a very diverse one. Most snails that live in saltwater respire using a gill or gills, a few species however have a lung, are intertidal, and are active only at low tide when they can move around in the air. These air-breathing species include false limpets in the family Siphonariidae and another group of false limpets in the family Trimusculidae.

Many (but not all) sea snails have an operculum.

The shell of *Syrinx aruanus* can be up to 91 cm long

Human uses

A number of species of sea snails are exploited by humans for food, including abalone, conch, limpets, whelks (such as the North American *Busycon* species and the North Atlantic *Buccinum undatum*) and periwinkles including *Littorina littorea*.

A number of species of edible whelks for sale at a fish market in Japan

The shells of sea snails are often found by humans as one kind of seashell that washes up on beaches. Because the shells of many sea snails are attractive and durable, they have been used by humans to make necklaces and other jewelry from prehistoric times to the current day.

The shells of a few species of large sea snails within the Vetigastropoda have a thick layer of nacre and have been exploited as a source of mother of pearl. Historically the button industry relied on these species for a number of years.

Use by other animals

The shells of sea snails are used for protection by many kinds of hermit crabs. A hermit crab carries the shell by grasping the central columella of the shell using claspers on the tip of its abdomen.

Definition

It is not always easy to decide whether some gastropods should be called sea snails. Some species that live in brackish water (such as certain neritids) can be listed as either freshwater snails or marine snails, and some species that live right at, or right above, the high tide level (for example species in the genus *Truncatella*), are sometimes considered to be sea snails and sometimes listed as land snails.

Taxonomy

2005 taxonomy

The following cladogram is an overview of the main clades of living gastropods based on the taxonomy of Bouchet & Rocroi (2005),[1] with taxa that contain saltwater or brackish water species marked in **boldface** (some of the highlighted taxa consist entirely of marine species, but some of them also contain freshwater or land species.)

A hermit crab which is occupying a shell of *Acanthina punctulata* has been disturbed, and has retracted into the shell, using its claws to bar the entrance in the same way that the snail used its operculum.

- Clade **Patellogastropoda**
- Clade **Vetigastropoda**
- Clade **Cocculiniformia**
- Clade **Neritimorpha**
  - Clade **Cycloneritimorpha**
- Clade **Caenogastropoda**
  - Informal group **Architaenioglossa**
  - Clade **Sorbeoconcha**
  - Clade **Hypsogastropoda**
    - Clade **Littorinimorpha**
    - Informal group **Ptenoglossa**
    - Clade **Neogastropoda**
- Clade **Heterobranchia**
  - Informal group **Lower Heterobranchia**
  - Informal group **Opisthobranchia**
    - Clade **Cephalaspidea**
    - Clade **Thecosomata**
    - Clade **Gymnosomata**
    - Clade **Aplysiomorpha**
    - Group **Acochlidiacea**
    - Clade **Sacoglossa**
    - Group **Cylindrobullida**

- Clade **Umbraculida**
- Clade **Nudipleura**
  - Clade **Pleurobranchomorpha**
  - Clade **Nudibranchia**
    - Clade **Euctenidiacea**
    - Clade **Dexiarchia**
      - Clade **Pseudoeuctenidiacea**
      - Clade **Cladobranchia**
        - Clade **Euarminida**
        - Clade **Dendronotida**
        - Clade **Aeolidida**
- Informal group Pulmonata
  - Informal group Basommatophora
  - Clade Eupulmonata
    - Clade **Systellommatophora**
    - Clade Stylommatophora
      - Clade Elasmognatha
      - Clade Orthurethra
      - Informal group Sigmurethra

Cultural references

In the animated American TV series *SpongeBob SquarePants*, the main character SpongeBob has a pet sea snail called Gary. Gary makes a "meow" vocalization like a cat. The character's eyes are well-developed and colorful, similar to the eyes of species in the sea snail family Strombidae.

See also

- Freshwater snail
- Land snail
- Sea slug
- Slug

References

[1] Bouchet P., Rocroi J.-P., Frýda J., Hausdorf B., Ponder W., Valdés Á. & Warén A. (2005). "[[Taxonomy of the Gastropoda (Bouchet & Rocroi, 2005)|Classification and nomenclator of gastropod families (http://www.archive.org/details/malacologia47122005inst)]"]. *Malacologia: International Journal of Malacology* (Hackenheim, Germany: ConchBooks) **47** (1-2): 1–397. ISBN 3925919724. ISSN 0076-2997. .

Article Sources and Contributors

Pikachuwashere, Pill, Pinky sl, Pit, Pizza12456, Pleasantville, Plumbago, Pmineault, Popefauvexxiii, Potatoepoptarts, Potatoswatter, Prashanthns, Princesspeej, Pritha 123, Pthag, Puffin, Pyrospirit, Qwertyuiopasdfghjkl12345, Qwyrxian, R'n'B, RadiantRay, Rayshade, Rc pimp, Recognizance, Red, Res2216firestar, RexNL, Rgamble, Rhobite, Rjwilmsi, Rmcgrath, Romanempire, Rooster Man 3, Roseindela, Roux, Rror, Rtyq2, Rzman11, SMP, SP-KP, Sam Blacketer, Samak47, Sampi, SchfiftyThree, Scjessey, SebastianHelm, Secretlondon, Seddon, Shadowlynk, Shaniece622, Sheeana, Shimgray, SifynCN, Siim, Skittleys, Slon02, SmartGuy, Smith609, Snek01, SnoozingInTheLemonGrove, Snowmanradio, Snowshovel7, Sonicblue4, Spaully, Spellmaster, Sprachpfleger, Stefan, Stemonitis, Steven J. Anderson, Stevenyang30, Storkk, Storm Rider, Strykers10, Studerby, Stwalkerster, Substatique, TUF-KAT, Taborgate, Tally Youngblood, Taollan82, Tarquin, TeaDrinker, TedE, TedPavlic, Template namespace initialisation script, That Guy, From That Show!, The Hokkaido Crow, The Mysterious X, The bellman, TheLimbicOne, Them Guys, Thingg, Thisisborin9, Tide rolls, Tigershrike, Tim1357, Tomi, Travelbird, TrickyP, Ucucha, Ugulee, Uncle Dick, UtherSRG, Vandal B, Velella, Versus22, Vina, WHeimbigner, WadeSimMiser, Waggers, Wavelength, Wayne Slam, Webridge, West Brom 4ever, Wham Bam Rock II, Wikidenizen, Wilford Nusser, Willhowell, Willking1979, Wimt, Wisdom89, Withaneduh, Wloveral, Wmahan, Yamamoto Ichiro, Zazazen, \\\\\\pwarg\\\\\\, 895 anonymous edits

Sea_snail *Source*: http://en.wikipedia.org/w/index.php?title=Sea_snail *Contributors*: Allen3, Anna Frodesiak, BD2412, Carlossuarez46, Crzrussian, Dan653, Enochlau, FerRacimo, Invertzoo, JForget, JNW, January, Masonisaskuxx, Nicolasheyes, OlEnglish, Omagomagom, PotatoSamurai, Redglasses, Reywas92, Rsrikanth05, Smalljim, Snek01, Some jerk on the Internet, The Thing That Should Not Be, Tide rolls, Unreal7, Xyzzyplugh, Yungchivo, 28 anonymous edits

Image Sources, Licenses and Contributors

File:Biological_classification_L_Pengo_vflip.svg *Source*: http://en.wikipedia.org/w/index.php?title=File:Biological_classification_L_Pengo_vflip.svg *License*: unknown *Contributors*: Adrignola, ArnoLagrange, Nisetpdajsankha, Pavel55, Pengo

File:Phylloscopus trochiloides NAUMANN.jpg *Source*: http://en.wikipedia.org/w/index.php?title=File:Phylloscopus_trochiloides_NAUMANN.jpg *License*: unknown *Contributors*: Anniolek, Kilom691, Nicke L, Red devil 666

File:Undiscovered species chart.png *Source*: http://en.wikipedia.org/w/index.php?title=File:Undiscovered_species_chart.png *License*: unknown *Contributors*: User:KVDP

File:Carolus Linnaeus (cleaned up version).jpg *Source*: http://en.wikipedia.org/w/index.php?title=File:Carolus_Linnaeus_(cleaned_up_version).jpg *License*: unknown *Contributors*: Original painting by Alexander Roslin. Digitally improved by Greg L.

file:Common_nutmeg_12.jpg *Source*: http://en.wikipedia.org/w/index.php?title=File:Common_nutmeg_12.jpg *License*: unknown *Contributors*: Original uploader was Edwardtbabinski at en.wikipedia

File:Snail diagram-en edit1.svg *Source*: http://en.wikipedia.org/w/index.php?title=File:Snail_diagram-en_edit1.svg *License*: unknown *Contributors*: User:Al2, User:Jeff Dahl

File:Prosobranchia male.png *Source*: http://en.wikipedia.org/w/index.php?title=File:Prosobranchia_male.png *License*: unknown *Contributors*: User:snek01

File:Zonitoides nitidus drawing.svg *Source*: http://en.wikipedia.org/w/index.php?title=File:Zonitoides_nitidus_drawing.svg *License*: unknown *Contributors*: User:snek01

File:Vinogradski puz glava.jpg *Source*: http://en.wikipedia.org/w/index.php?title=File:Vinogradski_puz_glava.jpg *License*: unknown *Contributors*: User:Tsnena

File:Haliotis asinina trochophore.jpg *Source*: http://en.wikipedia.org/w/index.php?title=File:Haliotis_asinina_trochophore.jpg *License*: unknown *Contributors*: Daniel J Jackson, Gert Wörheide and Bernard M Degnan

File:Elysia timida mating.jpg *Source*: http://en.wikipedia.org/w/index.php?title=File:Elysia_timida_mating.jpg *License*: unknown *Contributors*: Valerie Schmitt, Nils Anthes and Nico K Michiels

File:JurassicMarineIsrael.JPG *Source*: http://en.wikipedia.org/w/index.php?title=File:JurassicMarineIsrael.JPG *License*: unknown *Contributors*: User:Wilson44691

File:Snail1web.jpg *Source*: http://en.wikipedia.org/w/index.php?title=File:Snail1web.jpg *License*: unknown *Contributors*: ComputerHotline, Jymm, Snek01, Überraschungsbilder

File:Slimaczek.jpg *Source*: http://en.wikipedia.org/w/index.php?title=File:Slimaczek.jpg *License*: unknown *Contributors*: User:Stako

File:Turritellatricarinata.jpg *Source*: http://en.wikipedia.org/w/index.php?title=File:Turritellatricarinata.jpg *License*: unknown *Contributors*: User:Wilson44691

File:Fulguropsis radula 01.JPG *Source*: http://en.wikipedia.org/w/index.php?title=File:Fulguropsis_radula_01.JPG *License*: unknown *Contributors*: H. Zell

File:Cypraea chinensis with partially extended mantle.jpg *Source*: http://en.wikipedia.org/w/index.php?title=File:Cypraea_chinensis_with_partially_extended_mantle.jpg *License*: unknown *Contributors*: Bricktop, GrahamBould, JoJan, Kilom691

Image:Mollusc generalized.png *Source*: http://en.wikipedia.org/w/index.php?title=File:Mollusc_generalized.png *License*: unknown *Contributors*: Philcha (talk) 08:04, 9 October 2008 (UTC). Original uploader was Philcha at en.wikipedia

Image:Snail radula working.png *Source*: http://en.wikipedia.org/w/index.php?title=File:Snail_radula_working.png *License*: unknown *Contributors*: User:Debivort. Original uploader was Philcha at en.wikipedia

File:Gastropod nervous system.gif *Source*: http://en.wikipedia.org/w/index.php?title=File:Gastropod_nervous_system.gif *License*: unknown *Contributors*: User:Anaxial

Image:Trochophore larva 01.png *Source*: http://en.wikipedia.org/w/index.php?title=File:Trochophore_larva_01.png *License*: unknown *Contributors*: Wlodzimierz. Original uploader was Philcha at en.wikipedia

Image:Yochelcionella water flow.png *Source*: http://en.wikipedia.org/w/index.php?title=File:Yochelcionella_water_flow.png *License*: unknown *Contributors*: Philcha (talk) Original uploader was Philcha at en.wikipedia

Image:Neptunea despecta.jpg *Source*: http://en.wikipedia.org/w/index.php?title=File:Neptunea_despecta.jpg *License*: unknown *Contributors*: Nyst

Image:Nautiloid septa n siphuncle 01.png *Source*: http://en.wikipedia.org/w/index.php?title=File:Nautiloid_septa_n_siphuncle_01.png *License*: unknown *Contributors*: Original uploader was Philcha at en.wikipedia Later version(s) were uploaded by Smith609 at en.wikipedia.

Image:Pearl farm (Seram, Indonesia).jpg *Source*: http://en.wikipedia.org/w/index.php?title=File:Pearl_farm_(Seram,_Indonesia).jpg *License*: unknown *Contributors*: User:Mark Richards

Image:Meister von San Vitale in Ravenna 004.jpg *Source*: http://en.wikipedia.org/w/index.php?title=File:Meister_von_San_Vitale_in_Ravenna_004.jpg *License*: unknown *Contributors*: -

Image:Hapalochlaena lunulata.JPG *Source*: http://en.wikipedia.org/w/index.php?title=File:Hapalochlaena_lunulata.JPG *License*: unknown *Contributors*: Jens Petersen

Image:Textile cone.JPG *Source*: http://en.wikipedia.org/w/index.php?title=File:Textile_cone.JPG *License*: unknown *Contributors*: Ausxan, InverseHypercube, Rling, Valérie75, 1 anonymous edits

Image:Schistosomiasis itch.jpeg *Source*: http://en.wikipedia.org/w/index.php?title=File:Schistosomiasis_itch.jpeg *License*: unknown *Contributors*: Abanima, Liftarn, Salvadorjo

File:Epitonium billeanum (Wentletrap).jpg *Source*: http://en.wikipedia.org/w/index.php?title=File:Epitonium_billeanum_(Wentletrap).jpg *License*: unknown *Contributors*: User:Nhobgood

File:Common limpets1.jpg *Source*: http://en.wikipedia.org/w/index.php?title=File:Common_limpets1.jpg *License*: unknown *Contributors*: User:Tango22

File:Syrinx aruanus shell.jpg *Source*: http://en.wikipedia.org/w/index.php?title=File:Syrinx_aruanus_shell.jpg *License*: unknown *Contributors*: Stephaniedancer

File:Whelks at a fish market in Japan.jpg *Source*: http://en.wikipedia.org/w/index.php?title=File:Whelks_at_a_fish_market_in_Japan.jpg *License*: unknown *Contributors*: me

File:Acanthina punctulata.jpg *Source*: http://en.wikipedia.org/w/index.php?title=File:Acanthina_punctulata.jpg *License*: unknown *Contributors*: Jerry Kirkhart